Integrated
Automotive
Safety Handbook

Other SAE books of interest:

**Occupant Protection and Automobile
Safety in the U.S. since 1900**
(Product Code: R-404)

**Vehicle Accident Analysis and Reconstruction
Methods, Second Edition**
(Product Code: R-397)

**Forensic Biomechanics: Using Medical
Records to Study Injury Mechanisms**
(Product Code: R-379)

For more information or to order a book, contact
SAE International
400 Commonwealth Drive,
Warrendale, PA 15096-0001, USA;

phone 877-606-7323 (U.S.A. and Canada only)
or 724-776-4970 (outside U.S.A. and Canada);
fax 724-776-0790;

emailCustomerService@sae.org;
website http://books.sae.org.

Integrated Automotive Safety Handbook

Ulrich Seiffert and Mark Gonter

Warrendale, Pennsylvania, USA

400 Commonwealth Drive
Warrendale, PA 15096-0001 USA

E-mail: CustomerService@sae.org
Phone: 877-606-7323 (*inside USA and Canada*)
          724-776-4970 (*outside USA*)
Fax: 724-776-0790

ISBN 978-0-7680-6437-7
SAE Order Number R-407
DOI 10.4271/R-407

Library of Congress Cataloging-in-Publication Data
Seiffert, Ulrich.
   Integrated automotive safety handbook / Ulrich Seiffert and Mark Gonter.
       pages cm
   "SAE order number R-407"—Title page verso.
   Includes bibliographical references.
   ISBN 978-0-7680-6437-7
1. Automobiles—Safety measures.   2. Traffic safety.   3. Automobiles—Technological innovations.   I. Gonter, Mark. II. Title.
   TL242.S43 2013
   629.222028'9—dc23
                                                          2013018537

Information contained in this work has been obtained by SAE International from sources believed to be reliable. However, neither SAE International nor its authors guarantee the accuracy or completeness of any information published herein and neither SAE International nor its authors shall be responsible for any errors, omissions, or damages arising out of use of this information. This work is published with the understanding that SAE International and its authors are supplying information, but are not attempting to render engineering or other professional services. If such services are required, the assistance of an appropriate professional should be sought.

To purchase bulk quantities, please contact:

SAE Customer Service
E-mail: CustomerService@sae.org
Phone: 877-606-7323 (*inside USA and Canada*)
          724-776-4970 (*outside USA*)
Fax: 724-776-0790

Visit the SAE International Bookstore at

books.sae.org

Table of Contents

Preface

This book describes all areas of vehicle safety: accident avoidance, pre-crash technologies, mitigation of injuries, and post-crash technologies. Special attention is given to driver assistance systems and to compatibility between vehicles in car-to-car crashes as well as pedestrian protection. Several countries have achieved a high level of vehicle safety; however, more than 1.2 million fatalities still occur each year on roadways worldwide. These metrics indicate a continuing need to improve vehicle and road safety.

New technologies in sensors and electronic control units and the growing knowledge of car-to-car and car-to-infrastructure technologies have fused the previously separate areas of accident avoidance (popularly known as "active safety") and injury mitigation (popularly known as "passive safety") into the newer concept of "integrated vehicle safety." This new approach represents a further step in reducing accident rates. In this book we detail a significant number of integrated vehicle safety solutions.

We hope that this book will be useful for those who are interested in the complex field of automotive safety. In particular, experts from industry and academia, as well as students, can learn new details of vehicle safety engineering within the broad perspective of vehicle safety today.

Both of us have many years of experience in the field of vehicle safety engineering in both industrial research and development (R&D) and as associated lecturers at the University of Braunschweig, Germany. We thank the many people who have supported the creation of this book, especially from Audi AG, Daimler AG, and Volkswagen AG.

The views and opinions expressed in this book are those of the authors and not necessarily those of any academic institution or other entity.

—Ulrich Seiffert and Mark Gonter

Chapter 1
The Need to Increase Road Safety

1.1 Introduction

After years of continuous vehicle development and traffic infrastructure improvements, worldwide mobility requirements are changing rapidly. There are various reasons for these changes including:

- Increase in the number of megacities
- Increase in mobility needs throughout the world, especially in China and India
- Introduction of new technologies including car-to-car and car-to-infrastructure, intelligent vehicles, electrical drive line, and fuel cells
- Introduction of several energy sources: crude oil, electrical power, bio-fuel

Many different requirements influence the further development and production of cars, for example, lower emissions in general but more specifically carbon dioxide (CO_2) emissions, higher levels of road safety, integrated mobility concepts, better vehicle quality, lower vehicle cost, new owner concepts such as car sharing and, last but not least, innovative concepts for vehicle weight reduction and intelligent vehicles. Infrastructure and vehicle safety will have even greater importance due to the high number of fatal injuries in road traffic accidents. Today worldwide, there are more than 1.2 million road accident fatalities annually, which leads to increased accident reduction activities, not only on the national but also on the international level.

Even the United Nations has initiated worldwide programs to reverse the negative trend. At the International Federation of Automotive Engineering Societies (FISITA) summit in November 2010 [1-1], one hundred senior technical executives from leading automotive companies joined with experts and policymakers from around the world to address the issue of global traffic safety. They raised the following questions and worked out the appropriate answers.

Question: Which should come first, improvements in infrastructure or vehicle technology?

Answers:

- Both improvements to infrastructure safety and vehicle safety are needed at the same time.
- Improvements in infrastructure safety are lagging behind improvements in vehicle safety in the developing world.
- Vehicle safety is being globally driven and is maturing.
- National opportunities to influence vehicle safety will be small and incremental.
- National opportunities to influence infrastructure safety will be greater, but will still be driven by universal safety design principles.
- There is considerable potential in the deepening dialogue between automotive safety engineers and road infrastructure safety engineers, and this FISITA summit is a good start, as is the emerging dialogue with the World Road Association PIARC.
- The third partner in this dialogue will be governments that manage their transport sectors.

Question: How much farther can active and passive vehicle safety technologies really take us?

Answers:

- Vehicle technology is not at its limit yet; new technologies will continue to increase road safety in the future.
- Installation rates of modern assistance systems need to be raised and awareness must be increased at both the end-user level (e.g., by national campaigns) and with original equipment manufacturing (OEM) personnel at the point of sale.
- Harmonized regulatory requirements, both functional and legal, in all markets will enhance the development and rollout of new technology.
- Quick wins, for example, separating vulnerable road users from vehicles, have great potential for road safety in developing countries.
- Data driven evaluations of the problem are essential before finding solutions; therefore, internationally harmonized in-depth accident research is needed.
- Seat belts, airbags, and electronic stability control (ESC) may comprise a basic set of required safety equipment for all markets; optional "musts" depend on cost/benefit evaluations based on above approach.

Question: Is the pace of modern living compatible with good traffic safety?

Answers:

- Combining the strengths of both worlds (human and technology) can maximize the benefit and minimize the risk of enhanced safety.
- Although human behavior is the main causal factor for most accidents, technology performance must be proven superior before it replaces the driver. To

achieve this, development of standardized effectiveness assessment methods are required.

- Changing human risk and safety perception is complex; therefore, legislation and education are needed. Education is one important approach to encourage awareness regarding traffic safety.

- Legislation/enforcement should always be the first countermeasure in changing human behavior; for example, standards can be developed for and compliance increased by education.

- Governments have to be more involved in the process to reach the "vision zero," especially in developing countries.

- How fast these recommendations can become reality depends very much on the common activities in the various countries.

In May 2011, the United Nations officially launched its "Decade of Action for Road Safety," an unprecedented global effort which seeks to save five million lives over a ten-year period. Overseeing the planning and execution of the Decade of Action is the responsibility of the United Nations Road Safety Collaboration (UNRSC), a partnership of governments, international agencies, civil society organizations, and private companies from more than 100 countries. The World Health Organization is the secretariat of the UNRSC.

The UNSRC is currently developing a plan for the Decade of Action based on five pillars:

- Building road safety management capacity
- Improving the safety of road infrastructure and broader transport networks
- Further developing the safety of vehicles
- Enhancing the behavior of road users
- Improving post-crash care

In November 2011, at its meeting in Geneva, the UNRSC voted to accept FISITA as a member organization; therefore, the expertise of the automotive engineers can contribute to the necessary improvements especially as they relate to safe vehicles.

1.2 Definitions

Vehicle and road safety can be defined as shown in Figure 1.1. Very early definitions are described by Wilfert and Seiffert [1-2, 1-3].

The following terms are commonly used in the automotive field.

- **Integrated vehicle safety:** combination of accident avoidance and mitigation of injuries

- **Accident avoidance:** (Active safety) all measures that are used to avoid accidents

- **Mitigation of injuries:** (Passive safety) all measures that help to reduce injuries during and after the crash

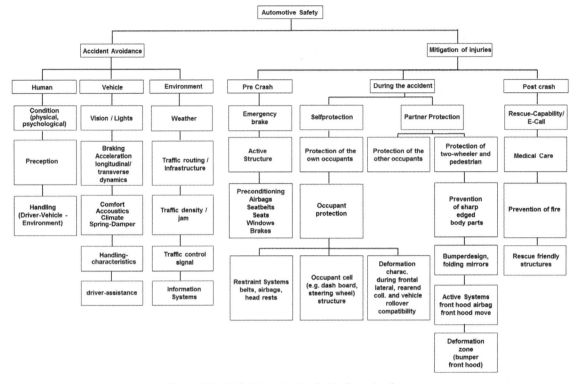

Figure 1.1: Definitions in the field of road safety.

- **Pre-crash:** vehicle systems that attempt to minimize accident consequences by "reversible" and "nonreversible measures" when the accident is unavoidable

- **External safety:** design of the external parts of the vehicle to reduce injuries in the event of a collision

- **Interior safety:** design of the interior parts of the vehicle to help prevent additional occupant injuries

- **Restraint systems:** vehicle components (e.g., seat belts, airbags, head restraints) that specifically influence the relative movement of occupants during an accident

- **Smart restraints, sensors, and actuators:** apply to occupant detection and pre-crash evaluation

- **Primary collision:** collision of the vehicle with another obstacle

- **Secondary collision:** collision of the occupant with vehicle components

The principal cause of accidents is still the human being. Although accident causes vary greatly, the same driver very often also shows a different accident risk with different vehicle models. Numerous driving assistance systems will further reduce the number of accidents assuming that drivers are driving without the influence of alcohol, medications, and illegal drugs; are in good physical condition; are in a comfortable vehicle with good ergonomic design; and have sufficient lighting and a good field of view. Driving assistance systems include brakes, powertrain, anti-lock braking systems (ABS), yaw-moment control, brake assist, and stability and insensitivity to crosswinds as well as communication and information systems. Figure 1.2 shows the reasons for accidents, related to driver error [1-4], based on a study done by the accidents research group of the German automobile club, AllgemeinerDeutscherAutomobil-Club (ADAC).

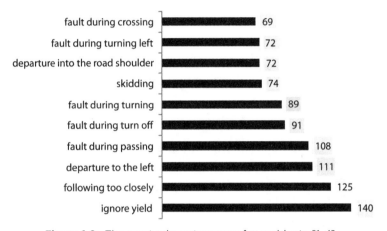

Figure 1.2: The most relevant reasons for accidents [1-4].

The number of driver assistance systems is increasing. Today the following features are available:

- Adaptive cruise control
- Emergency brake assistant
- Lane change/blind spot assistant
- Drowsiness assistant
- Traffic-sign recognition
- High beam assistance, curve lighting
- Pedestrian and animal recognition
- Pedestrian protection through emergency braking
- Electronic Stability Control (ESC)
- Tire pressure monitoring system
- Hill start assistant
- Park assistant

1.3 Driving forces for increased vehicle safety

Many factors have helped to reduce the number of accidents and their consequences including the following examples:

- Increasing consumer and public demand for road safety
- Technological competition between vehicle manufacturers and suppliers
- Legislation introduced by governments
- Product liability
- Results of accident analysis
- Consumer information
- Automotive press
- Insurance companies and related organizations

Details on some of these factors are provided in the following sections.

1.3.1 Legislation

The different government legislative institutions that are responsible for traffic safety combine the three elements: traffic routing (e.g., road design, traffic flow, and traffic signals), the education of traffic participants (e.g., drivers, pedestrians, and school children), and vehicle performance.

Legislation certainly has a strong influence on safety. In 1832, England introduced accident performance requirements for a steam-powered bus. In 1909, with the development of the first automobiles in Germany, a law covering vehicle liability was proposed. The rule-making activities in the United States (U.S.) during the mid-1960s and the International Experimental Safety Vehicle Conference, where safety became a competitive item for medical and technical scientists, legislators, and automotive engineers had a strong and positive influence on legislation.

A second effect was the activity initiated by consumer advocates such as Ralph Nader in the U. S., as well as the increasing number of consumer information reports with respect to the performance of vehicles in the field and in accident simulation tests, such as the crashworthiness rating reports.

It is most important, that requirements do not mandate design but rather specify performance criteria that must be met through defined, objective tests. This approach encourages engineering creativity and the competition of different concepts and ideas for a place in the design.

With respect to automotive legislation on a worldwide basis, there are different starting points for different countries. For example, in the U. S. the responsible government agencies were of the opinion that traffic participants, especially drivers, could be educated only to a limited extent; vehicle occupants must thus be protected in the event of an accident caused by the vehicle design. Measures for accident avoidance were assigned a lower priority. Contrary to the U.S. highway safety philosophy, European legislators placed much more responsibility on the driver because their focus was the prevention of accidents.

In the beginning and until the mid-1960s, the number of accidents with severe and fatal injuries reached such a high level that the U.S. government defined comprehensive requirements for the safety and crash performance of automobiles in November 1966, with the National Traffic and Motor Vehicle Safety Act of 1966. Europe and other countries followed suit. Meanwhile, more than 100 requirements have been introduced throughout the world.

In Europe, the rule-making process is as follows. Vehicles that are able to drive faster than 3.7 mph (6 km/h) and can be driven on public roads must fulfill national requirements or, if possible, apply for a European Economic Community (EEC)-type approval. If the vehicle manufacturer applies for an EEC-type approval, it must also prove that the manufacturing processes employ a suitable quality control system to assure the conformity of production (COP) to the specifications that were the subject of type approval. A typical example is Germany. Table 1.1 shows the accident avoidance and injury minimization requirements for passenger vehicles with up to eight seats for Europe EEC, Germany (StVZO), and the United Nations Economic Commission for Europe (ECE). Table 1.2 shows the relevant standards in the U S.

Table 1.1: Legal requirements in Europe			
vehicle category	EC-Directive	ECE-Regulation	StVZO Road Traffic Licensing Regulation
Requirements for Accident Avoidance in Europe (Active Safety)			
Steering equipment	70/311EWG	R 79	§ 38
Brake-systems	71/320/EWG	R 13-H	§ 41
Replacement brake pads/ shoes	71/320/EWG	R 90	§ 22
Equipment for acoustic signal	70/388/EWG	R 28	§ 55
Field of view	77/649/EWG	—	§ 35b
Defrosting and defogging for glazing	VO (EG) 672/2010	—	§ 35b
Windshield wipers and washers	VO (EG) 1008/2010	—	§ 40
Rearview mirrors	2003/97/EG	R 46	§ 56
Heaters (engine waste heat)	2001/56/EG	R 122	§ 35c
Installation of lighting and lighting-signaling devices	76/756/EWG	R 48	§ 49a, 53a
Reflex reflector	76/757/EWG	R 3	§ 53

(continues)

Table 1.1: Legal requirements in Europe (Continued)			
vehicle category	EC-Directive	ECE-Regulation	StVZO Road Traffic Licensing Regulation
Requirements for Accident Avoidance in Europe (Active Safety) (Continued)			
Side marker lamps, rear and brake lamps	76/758/EWG	R 7	§ 51, 51b, 53
Side marker lamps	76/758/EWG	R 91	§ 51a
Turn signal lamps	76/759/EWG	R 6	§ 54
Headlamp for high beam and/or low beam	76/761/EWG	R 1, 8, 20, 112, 113	§ 50
Gaseous discharge head lamps and their light sources	— —	R 98 R 99	— —
Adaptive front beams	—	R 123	—
Fog lamps	76/762/EWG	R 19	§ 52
Turn lamps	—	R 119	—
Rear fog lamps	77/538/EWG	R 38	§ 53d
Rear drive lamps	77/539/EWG	R 23	§ 52a
Parking lamps	77/540/EWG	R 77	§ 51c
Rear registration plate illumination device	76/760/EWG	R 4	§ 60
Reverse gear and speed meter equipment	75/443/EWG	R 39	§ 39,57
Interior equipment (symbols, warning light)	78/316/EWG	R 121	§ 30
Wheel covers	VO (EG) 1009/2010	—	§ 36a
Tire tread depth	89/459/EWG	—	§ 36
Tire and tire mounting	92/23/EWG	R 30	§ 36
Tires and wet road performance	—	R 117	—
Reserve Tire	—	R 124	—
Towing capacity, hitch vertical load	92/21/EWG	—	§ 42,44
Towing equipment, trailer hitch	94/20/EG	R 55	§ 43
Pedal arrangement	—	R 35	§ 30
Requirements for mitigation of injuries in Europe(Passive Safety)			
Interior fittings (protruding elements)	74/60/EWG	R 21	§ 30
Steering mechanism (behavior in an impact)	74/297/EWG	R 12	§ 38
Frontal impact, occupant protection	96/79/EG	R 94	§ —

Table 1.1: Legal requirements in Europe (Continued)			
vehicle category	EC-Directive	ECE-Regulation	StVZO Road Traffic Licensing Regulation
Side impact, occupant protection	96/27/EG	R 95	§ —
Seat-belt anchorages	76/115/EWG	R 14	§ 35a
Seat belts and restraint systems	77/541/EWG	R 16, R 44	§ 22a, 35a
Seat, seat anchorages, head restraints	74/408/EWG	R 17, R 25	§ 35a
Head restraints	78/932/EWG	R 17, R 25	§ 35a
External protection	74/483/EWG	R 26	§ 30c
Fuel tanks and rear underride protection	70/221/EWG	R 58	§ 47-47c
Liquefied petroleum fuel systems	—	R 67, R 110, R 115	§ 41a, 45, 47
H_2-driven vehicle	VO (EG) 79/2009	—	—
Doors (locks and hinges)	70/387/EWG	R 11	§ 35e
Front and rear bumpers	—	R 42	—
Rear-end collisions (not applicable to Germany)	—	R 32	—
Safety glazing	92/22/EWG	R 43	§ 40
Electric propulsion (safety)	—	R 100	§ 62
Pedestrian protection	VO (EG) 78/2009	—	—
Airbags	—	R 114	—
Luggage separation net	—	R 126	—

Based on the 1958 agreement covering common rules for the approval of parts and vehicles, the United Nations ECE is working to harmonize the different requirements throughout the world. With the amendments of October 16, 1995, the working title became the "New agreement concerning the adoption of uniform technical prescriptions for wheeled vehicles, equipment, and parts that are used in road vehicles . . ." Application for membership is on a voluntary basis. This means that countries that are members of the United Nations but are not members of the EEC can join this commission.

Since March 1998, and November 1999, the EEC and Japan, respectively, became members of this commission. More than 100 rules are now in effect, some of which could be used as part of the EEC-type approval. Additional actions to minimize trade barriers among the countries and markets in the U.S., Japan, and Europe commenced with the Transatlantic Economy Dialog whose goal is to improve trade relations among the dialog's partners.

Table 1.2: Relevant Rules In The United States	
FMVSS	Contents
101	Controls and displays
102	Transmission shift lever sequence, starter interlock, and transmission braking effect
103	Windshield defrosting and defogging systems
104	Windshield wiping and washing systems
105	Hydraulic and electric brake systems
106	Brake hoses
108	Lamps, reflective devices, and associated equipment
109	New pneumatic tries
110	Tire selection and rims
111	Rearview mirrors
113	Hood latch system
114	Theft protection
116	Motor vehicle brake fluids
118	Power-operated window, partition, and roof panel systems
119	New pneumatic tries for vehicles other than passenger cars
120	Tire selection and rims for motor vehicles other than passenger cars
121	Air brake systems
124	Accelerator control systems
129	New non-pneumatic tires for passenger cars
135	Passenger car brake system
138	Tire monitoring systems
201	Occupant protection in interior impact
202	Head restraints
203	Impact protection for the driver from the steering control system
204	Steering control rearward displacement
205	Glazing materials
206	Door locks and door retention components
207	Seating systems
208	Occupant crash protection
209	Seat-belt assemblies
210	Seat-belt assembly anchorages
212	Windshield mounting

Table 1.2: Relevant Rules In The United States (Continued)	
FMVSS	Contents
213	Child restraint systems
214	Side impact protection
216	Roof crush resistance
219	Windshield zone intrusion
301	Fuel system integrity
302	Flammability or interior materials
303	Fuel system integrity of compressed natural gas vehicles
304	Compressed natural gas fuel container integrity
305	Electrical powered vehicles, electrolyte spillage and electrical shock protection

In the meantime, the following internet addresses provide information about the regulations in the different countries.

UN ECE:

http://unece.org7trans/main/welcwp29.htm

U.S.:

http://www.nhtsa.dot.gov/ and www.regulations.gov

Canada:

http://www.tc.gc.ca/acts-regulations/menu.htm

Australia:

http://www.infrastructure.gov.au/roads/motor/design/adr_online.aspx

India:

http://www.siamindia.com/scripts/rules.aspx

Japan:

http://www.nasva.go.jp

Korea:

http://english.koti.re.kr/

Brazil:

http://www.denatran.gov.br/resolucoes.htm

1.3.2 Competition

As early as the mid-1950s, research to improve vehicle performance during accidents began at universities and within the automotive industry. Professor Koeßler, Technical University of Braunschweig defined the task: a motor vehicle must transport humans and goods from A to B as safely, as quickly, and as comfortably as possible [1-2]. Mr. Barényi [1-5] received a patent related to the performance of a vehicle during the crash phase, Figure 1.3.

As a result of world-wide activities over the last decades, automobile manufacturers have contributed very significantly to the safety level and crash performance of their vehicles. The safety level for accident avoidance and injury mitigation is an important part of the research and development activities and is reinforced by the desire to make the vehicle as safe as possible.

Furthermore, safety legislation in the U.S. and the Experimental Safety Program (ESV), an international program to exchange views on vehicle safety organized by various government agencies worldwide, have encouraged and accelerated technological progress and competition among vehicle manufacturers. This, in turn, has contributed significantly to the positive results that have been achieved to date. Insurance organizations in the United States and elsewhere, for example the Insurance Institute for Highway Safety in the United States (IIHS); the consumer tests performed by government agencies, technical organizations, and the automotive media; and news and television reports, not to mention products liability, have all contributed in their own way to technological progress. The interest of the car buyer in vehicle safety is also increasing and many consumers are using internet information on vehicle safety before they buy a new car.

1.3.3 Consumer information

Aside from legislative activities, worldwide consumer information has a major impact on the decisions made by car buyers. When consumer tests first began, the main focus was on crash tests for the protection of adult vehicle occupants. At a later stage, child occupant protection and pedestrian protection followed. Accident avoidance measures became increasingly important; safety assistance systems like safety belt reminders, speed limitation devices, and the installation of ESC contributed positively under vehicle rating programs carried out by various organizations. Figures 1.4 and 1.5 provide an overview of the various tests in different countries. These statistics are part of a service offered by Carhs [1-6] and demonstrate that the NCAP ratings are taken very seriously throughout the world. For the vehicle buyer the information is also available via the internet, and the intensive activities to improve NCAP-scores are ongoing.

Positive NCAP tests are not only used by vehicle manufacturers to market new vehicles; magazine and newspaper reports generate significant public interest.

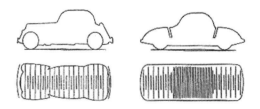

Figure 1.3: Patent drawing in 1952 by Béla Barényi [1-5].

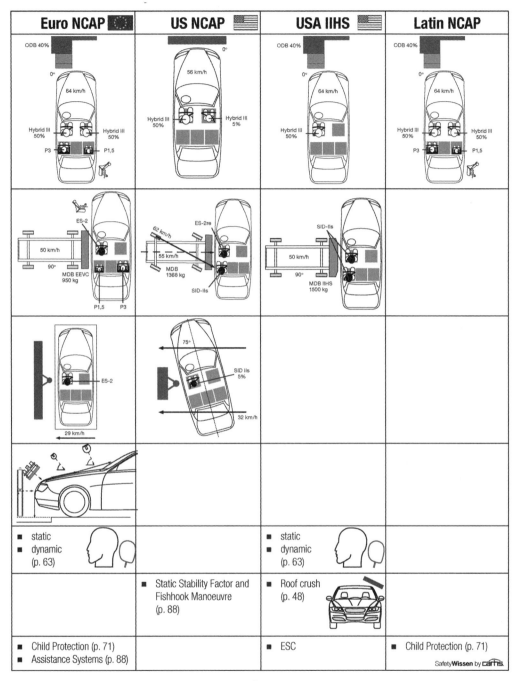

Figure 1.4: NCAP-Rating System for Europe, U.S., Latin America [1-6].

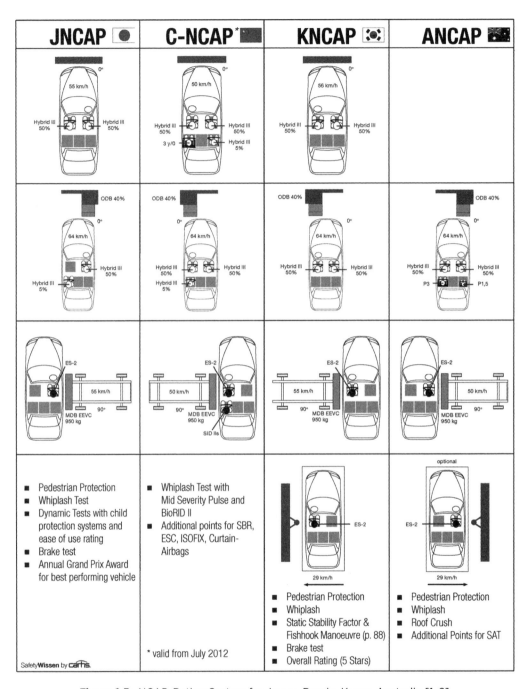

Figure 1.5: NCAP-Rating System for Japan, Russia, Korea, Australia [1-6].

NCAP criteria have become important factors in vehicle development and are subject to continuous upgrading and amending. The latest status of these initiatives is described by Professor Seeck [1-7]. It covers [1-6]:

- The new full width test
- Whiplash performance at the rear seats
- Emergency braking
- Child protection
- New overall balance criteria

1.3.4 Product liability

All vehicle manufacturers and sellers are responsible under the law for the performance of their vehicles. This not only means that legislative requirements must be met but also that performance in other areas must meet certain criteria and expectations—both written and unwritten. The high number of recalls carried out throughout the world demonstrates just how seriously vehicle manufacturers and importers take this responsibility. Product liability cases, often tried in the media, contribute to the fact that legal considerations, in addition to mandated technical requirements, exert a positive influence on the development of new vehicles.

1.4 References

1-1 FISITA World Automotive Summit 2010, Mainz Germany.

1-2 Seiffert, U. Fahrzeugsicherheit, VDI Verlag, Düsseldorf, 1992, ISBN 3-18-401264-6.

1-3 Wilfert, K. Entwicklungsmöglichkeiten im Automobil, ATZ 1973, S273-278.

1-4 ADAC-Motorwelt, Germany, page 20 08/2011.

1-5 Barényi, I.B. Das Prinzip des gestaltfesten Fahrerraums, Deutsches Patentamt 854.157 (1952).

1-6 Carhs, http:://www.carhs.de, carhsGmbH, Alzenau-Germany.

1-7 Seeck, A.NCAP-Tests, Consumer Tests, BAST, Graz Safety Update, September 2011.

Chapter 2
Accident Research

2.1 Introduction

Accident research and investigation has evolved and changed significantly during recent decades. In the beginning, data were used more for statistical evaluation to obtain a fuller understanding of the kinds of accidents that were occurring in the real world. Today, accident research and analysis is an integrated part of the research and development process in many companies throughout the world.

The primary reason for this change is the introduction of in-depth accident analysis and the close cooperation between science and industry. One good example is Germany. In 1979, an independent, multi-disciplinary working team from the medical school of the University of Hanover together with the Technical University of Berlin was formed and was financially supported by the German Federal Agency for Road Research (BAST). Today around 1000 accidents are documented and approximately 3000 data points are analyzed each year.

In 1999, these activities were significantly supplemented by the German In Depth Accident Study (GIDAS) conducted in cooperation with the Forschungsvereinigung Automobiltechnik (FAT) and BAST and supported by a second research team based at the University of Dresden. The criteria used in the study are a road accident where at least one person was injured.

Detailed data are recorded at the scene of an accident and supplemented with additional information regarding the accident scene, environmental factors, and data pertaining to the occupants and details of their injuries. These data permit subsequent, in-depth accident reconstruction. Supported by accident analysis teams fielded by vehicle manufacturers, this database has grown steadily over the years. Figure 2.1 depicts the procedure utilized by the Volkswagen accident research team.

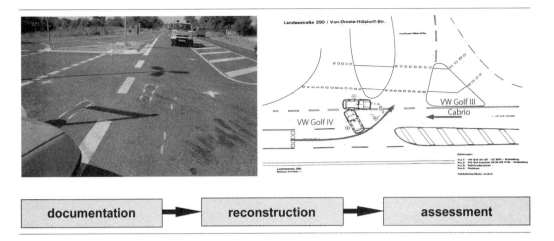

Figure 2.1: Principal procedure used by the Volkswagen accident research team [2-1].

In addition to GIDAS in Germany, similar in-depth accident research projects are in place in other countries including France, Great-Britain, Sweden, the U.S., Japan, and China. Unfortunately it is still not possible to utilize an identical data structure to enable data comparison and analysis of the real world accident experience in different countries and locations.

2.2   Accident data

There are many kinds of accident statistics available today including the largest database of documents created by IRTAD [2-2]. The following tables and figures highlight some of these accident data. For example, Table 2.1 shows data from countries that contribute to these statistics and how these statistics have evolved since 1970.

Figure 2.2 compares the deaths per billion vehicles—kilometers driven in 2009 or 2008, as indicated in several countries.

The difference in safety levels is significant. When viewed on a worldwide basis, we must note that there were more than 1.2 million deaths in 2009. The United Nations fears that traffic fatalities will increase to 2.4 million per year by 2030. Only 50 percent of these fatalities are vehicle occupants. The others are pedestrians, cyclists, and motorcyclists. More detailed information on traffic fatalities can be gleaned from the IRTAD database.

Figure 2.3 illustrates conditions in Germany by setting forth fatalities and injury crashes on an annual basis.

Other information that can be derived from the data includes the evaluation of fatality risk by age group. Figure 2.4 shows some comparable data for the U.S. [2-2].

Table 2.1: Traffic deaths per 100,000 inhabitants and per billion vehicle-kilometer 1970, 1980, 1990, 2000, and 2010

Country	Killed per 100 000 inhabitants					Killed per billion veh.-km				
	1970	1980	1990	2000	2010	1970	1980	1990	2000	2010
Argentina[a]	—	—	—	—	12.6	—	—	—	—	—
Australia	30.4	22.3	13.7	9.5	6.1	49.3	28.2	14.4	9.3	6.1
Austria	34.5	26.5	20.3	12.2	6.6	109.0	56.3	27.9	15.0	—
Belgium	31.8	24.3	19.9	14.4	8.8[b]	104.6[c]	50.0	28.1	16.4	9.6[b]
Cambodia[a]	—	—	—	3.4	12.7	—	—	—	—	—
Canada	23.8	22.7	14.9	9.4	6.6[b]	—	—	—	9.3	6.5[b]
Czech Republic	20.2	12.2	12.5	14.5	7.6	—	53.9	48.3	36.7	16.2
Denmark	24.6	13.5	12.4	9.3	4.6	50.5	25.0	17.3	10.7	5.6
Finland	22.9	11.6	13.1	7.7	5.1	—	20.6	16.3	8.5	5.1
France	32.6	25.4	19.8	13.7	6.4	90.4	44.0	25.7	15.1	7.1
Germany	27.3	19.3	14.0	9.1	4.5	—	37.3	20.0	11.3	5.2
Greece	12.5	15.0	20.2	18.7	11.1	—	—	—	—	—
Hungary	15.8	15.2	23.4	12.0	7.4	—	—	—	—	—
Iceland	9.8	11.0	9.5	11.5	2.5	—	26.5	14.9	13.8	2.6
Ireland	18.3	16.6	13.6	11.0	4.7	44.3	28.4	19.2	12.6	4.5
Israel	17.1	10.8	8.7	7.1	4.6	87.9	38.8	22.4	12.4	7.1
Italy	20.5	16.3	12.6	12.4	6.8	—	—	—	—	—
Japan	21.0	9.7	11.8	8.2	4.5	96.4	29.3	23.2	13.4	7.7[b]
Korea	11.0	17.0	33.1	21.8	11.3	—	—	—	49.5	18.7
Lithuania[a]	—	—	26.9	17.3	9.2	—	—	—	—	—
Luxemburg	—	27.0	18.8	17.5	6.4	—	—	—	—	—
Malaysia[a]	—	—	22.7	25.9	23.8	—	—	—	26.3	16.2
Netherlands	24.6	14.2	9.2	6.8	3.6	—	26.7	14.2	8.5	5.0[b]
New Zealand	23.0	18.8	21.4	12.1	8.6	—	—	—	13.6	9.4
Norway	14.6	8.9	7.8	7.6	4.2	41.7	19.3	12.0	10.5	4.9
Poland	10.6	16.8	19.2	16.3	10.2	—	—	—	—	—
Portugal	20.6	30.6	31.2	20.0	8.8	—	—	—	—	—
Serbia[a]	—	—	—	—	9.0	—	—	—	—	—
Slovenia	35.8	29.2	25.9	15.8	6.7	166.7	96.1	65.1	26.7	7.7
Spain	16.0	17.7	23.2	14.5	5.4	—	—	—	—	—
Sweden	16.3	10.2	9.1	6.7	2.8	35.3	16.4	12.0	8.5	3.2
Switzerland	26.6	19.2	13.9	8.3	4.2	56.5	30.9	18.6	10.6	5.3
United Kingdom	14.0	11.0	9.4	6.1	3.1	37.4[c]	21.9[c]	12.7[c]	7.3[c]	3.7[c]
United States	25.7	22.5	17.9	15.3	10.6	29.5	20.9	12.9	9.5	6.8

Death within 30 days. Police recorded data. Portugal: In 2010 there was a change of methodology in the calculation of the fatality data. p = provisional; a = accession country. Data are under review. b = 2009. c = Great Britain.

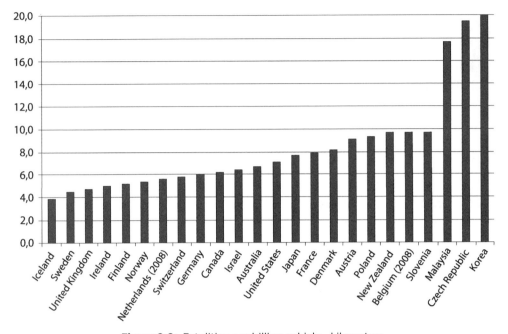

Figure 2.2: Fatalities per billion vehicle—kilometers.

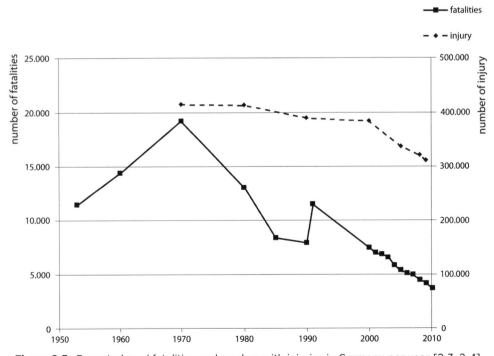

Figure 2.3: Reported road fatalities and crashes with injuries in Germany per year [2-3, 2-4].

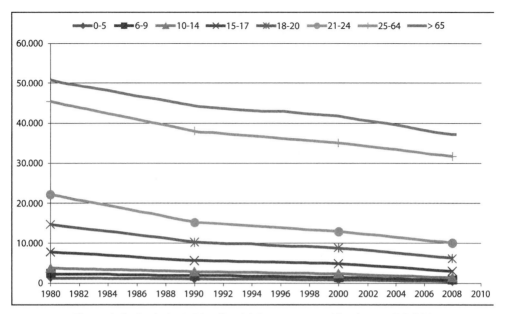

Figure 2.4: Evolution of fatality risk by age group (deaths per 100,000 population) 1990–2008.

The last example, Figures 2.5 and 2.6 [2-3], show the influence of road type on fatalities. Results for the U.S. are depicted in Figure 2.5, and results for Germany are depicted in Figure 2.6.

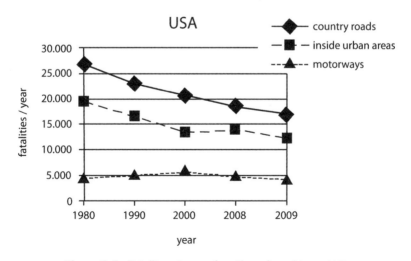

Figure 2.5: Fatality rate as a function of road type, U.S.

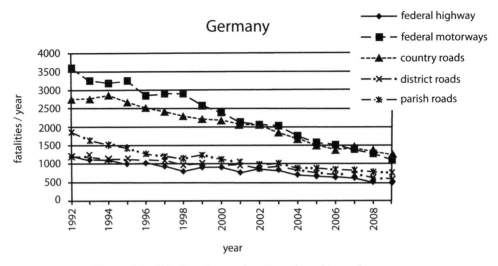

Figure 2.6: Fatality rate as a function of road type, Germany.

2.3 Application of accident research data

To be able to examine and compare the safety levels of different vehicles, various governmental authorities and consumer organizations conduct a wide variety of crash and other tests every year. The primary objective of these tests is to simulate a variety of real world accidents and optimize safety features and systems to help more effectively protect vehicle occupants and other participants in the traffic mix. Crash tests required by applicable safety standards are intended to assure a defined level of safety, while consumer testing seeks to document vehicle performance on a comparative basis and provide information used by the consumer for comparative shopping. It also encourages manufacturers to implement improvements as a part of their continuous development processes [2-5].

However, it is important to bear in mind that vehicle designs focused on compliance with laboratory test procedures tend to emphasize the vehicle itself and ignore other opportunities to improve vehicle and road and infrastructure safety. In addition, while concentrating on standardized testing, we must also determine whether increasing laboratory test requirements really improves vehicle performance and safety levels in real-world accident scenarios. The reduction of injuries through structured measures and advanced restraint systems could be demonstrated with the help of GIDAS-Project data [2-6]. On the basis of the accident analysis it is also possible to assess the effectiveness of safety systems. Based on the National Highway Traffic Safety Administration (NHTSA) analysis, the IIHS [2-7] reported, for example, that 12,713 lives were saved in the U. S. in 2009, and the use of helmets, mainly by motorcycle riders, saved 1483 lives in that same year. A ranking of different driver assistance systems and systems for injury mitigation with respect to anticipated or proven effectiveness is shown in Figure 2.7.

- Safety belts and safety belt pretensioner for the front seat
- Safety belts for the rear seats, child restraints
- Safety passenger cells
- ESC
- Lane Departure Alert
- Automatic Emergency Brake
- Drowsiness Monitoring
- Head airbag
- Driver airbag
- ABS, Brake Assist
- Front passenger airbag
- Lane Change Assist
- Side and knee airbags
- Pre-Safe systems
- Night Vision, Active Front Lighting

Figure 2.7: Ranking of safety features [2-8].

In recent research, Dr. Eichberger from the University of Graz and his team analyzed 43 different accident avoidance systems, evaluating their potential to reduce accidents [2-9, 2-10]. Further study using this methodology to analyze traffic and the accident environment is warranted. Evasive maneuver assistance, lane keeping assistance, emergency braking, and electronic stabilization systems, as well as autonomous driving may offer significant future traffic safety benefits.

Accident data use in other countries has also accelerated the introduction of new systems into the market. For example, ESC and features such as the forward collision avoidance system installed in the Volvo XC 60 midsize sport utility vehicle (SUV) [2-10] show great promise in increasing future traffic safety.

Evaluation of real-world accident simulations based on applicable test procedures and development of new technologies to improve vehicle and traffic safety is an ongoing process. For this reason it is necessary to gather accident data for subsequent detailed and multidisciplinary evaluation of large data files.

The evolution of vehicle accident data collection methodology is also evolving and continuously improving. For example, new safety features, such as ABS do not permit analysis and assessment of deceleration traces in the pre-collision phase. For the development of accident avoidance systems, reliable information about vehicle velocity, yaw and acceleration/deceleration, and driver reaction is required. These data could be obtained from standardized event data recorders (EDR). The introduction of these data recorders is still under discussion by the government agencies and vehicle manufacturers. This technology offers great potential to improve the quality of the accident data and support the development of effective accident avoidance systems [2-8].

Very often the question is raised: can the information obtained from accident data in one country can be transferred to other parts of the world? The answer to this question is given in the Figure 2.2. As long as the traffic situation in the various countries is so different, we will also find different priorities to address in reducing real-world accidents. Therefore, it is especially important that countries where the traffic density is rapidly increasing also establish accident investigation and research teams. One of the many positive examples of this trend is China [2-11], where the accident situation is showing positive trends. Generally, precise accident investigation analysis is a very important part of the safety engineering work.

2.4 References

2-1   Jungmichel, M., Stanzel, M., Zobel, R. Special Aspects in Accident Reconstruction in the Accident Investigation Department of Volkswagen, EVU—Tagung 2002, Portoroz, Slovenia.

2-2   IRTAD, International Road Traffic and Accident Database, www.irtad.net.

2-3   Verkehr in Zahlen 2010/2011 DIW Berlin, ISBN 978-3-87514-438-5.

2-4   Schwarz, T. Vergleich der Crashtestbedingungen für Personenkraftwagen mit dem realen Unfallgeschehen, Technische Universität Berlin, ISS—Fahrzeugtechnik, März 1999.

2-5   Becker, H., Sferco, R. Anwendung von Realunfalldaten in der Fahrzeugentwicklung am Beispiel des Frontairbags, Schriftenreihe der Bundesanstalt für Straßenwesen, Bergisch Gladbach, 2003.

2-6   IIHS Vol. 46 No. 7, Aug. 2011.

2-7   Schwarz Zobel. Handbuch Kraftfahrzeugtechnik, Vieweg—Teubner, 2011, ISBN 978-3-8348-1011-3.

2-8   Eichberger, A. et. al. Potenziale von Systemen der aktiven Sicherheit und Fahrerassistenz ATZ, 07-08,2011, 113, Jahrgang Wiesbaden.

2-9   Eichberger, A. Contribution to Primary, Secondary and Integrated Traffic Safety, Verlag Holzhausen GmbH, ISBN 978-3-85493-196-6.

2-10   IIHS Vol. 46 N06, July 2011.

2-11   Schöneburg, R. Neue Sicherheitstechnik und Fahrerassistenzsysteme bei Mercedes, ÖVK Wien-2011.

Chapter 3
Integrated Safety

3.1   Introduction

Safety has been an important issue ever since the development of the motorized vehicle. The major challenges for safe road traffic are infrastructure, education, and vehicle safety. Integrated safety includes both the driver and the roadside environment in the vehicle safety concept (Figure 3.1). The impact of car design on effective accident mitigation is the successful introduction of passive and active safety systems to avoid accidents and reduce or eliminate injuries. In the past, safety has been successfully developed further in two significant and largely independent areas. To date, the focus has been on passive safety. This has led to optimized structural design, for example, crumple zones and rigid passenger compartments, and the development of highly efficient restraint systems such as belt pre-tensioners, belt force limiters, and various airbag systems. Active systems are very powerful but highly dependent on sensors. Cars with active systems can feel, see, and communicate with other cars and with the environment.

The diversity of driver experience and the cultural diversity in different countries are very important issues to address in the layout of active systems. The existing infrastructure is an important determining and restrictive factor for the design. A driver's capability in handling the vehicle and dealing with human-machine interfaces is a challenge and a new field of operation for vehicle designers.

In addition to the technical benefit of such systems the question of cost effectiveness is a critical issue. The market must be willing to spend money for safety systems. Hence, the intelligent combination of both elements, passive safety design and driver assistance systems, will be a key for future safety enhancements in cars because of the brisk evolution of electronics. At the same time, additional developments in restraint systems, particularly intelligent restraint systems, can further exploit the potential of passive safety. Intelligent restraint systems can adapt their response to particular accident situations meaning that the accident severity

Figure 3.1: Integrated Safety includes the driver and the roadside environment in the vehicle safety concept.

and individual characteristics of the vehicle occupants can be better addressed (Figure 3.2).

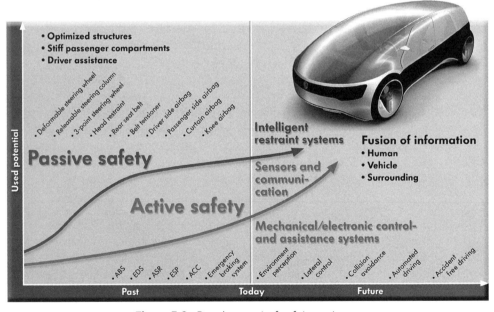

Figure 3.2: Development of safety systems.

By integrating active and passive safety into an integrated safety approach, it will be possible to greatly improve vehicle safety in the future, (Figure 3.3) [3-1, 3-2]. Integrating the systems at different system levels will result in new functional characteristics which will adaptively improve existing systems while making entirely new systems possible. By using shared sensors, integrated safety will make cost

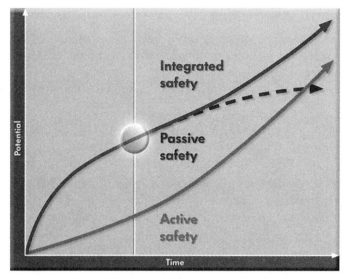

Figure 3.3: Potential for integrated safety.

effective implementation of various systems possible. As a result, the functional quality and availability of systems will improve.

Figure 3.4 shows the sub-areas of integrated safety. The classification is based on the temporal sequence of the functions, ranging from the initial intervention through an actual collision until the subsequent assistance during rescue. This represents the sequence of events in a typical accident. The objective must be to use optical, haptic, and/or audible warning signals to prompt the driver to avoid the accident. In this case, the driver is still more effective in many situations than current sensors based on his or her experience, visual capabilities, and interpretation of the situation. If the driver does not respond, the vehicle intervenes by braking and steering if necessary. In addition, shortly before the collision, when the accident can no longer be avoided, pre-crash measures take effect. System intervention requires high triggering qualities combined with a very low level of possible misinterpretations.

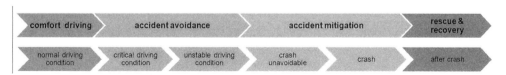

Figure 3.4: Sub-areas of integrated safety.

Combining different systems into an integrated safety system requires a modular situation-decision tool with different escalation levels governed by an integral protection concept. In the future, a holistic approach to safety functions during an accident and use of a superimposed decision tool will make it possible for the various individual functions to play their part in the total accident response function.

3.2 Accident avoidance

Not only do the assistance systems described below contribute to accident avoidance, but measures taken to improve comfort have a positive influence on the performance of the driver. These include, for example, air conditioning systems, lighting, and mirrors and windows that improve visibility as well as a balanced chassis.

The current status of development in mitigation and avoidance systems and their protection potential is shown in Figure 3.5. In this case, active safety can be divided into vehicle assistance (sensors that detect the vehicle status), driver assistance (environmental perception), and networked assistance (communication). The systems to assist the driver with environmental perception can be divided into systems for longitudinal guidance, lateral guidance, and night assistance.

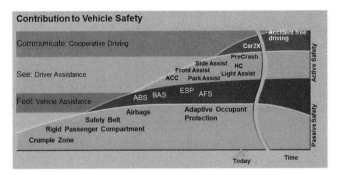

Figure 3.5: Potential of integrated vehicle safety systems [3-3].

In the future vehicles and infrastructure will communicate with each other, meaning that the potential for safety will further increase. Additional innovative development of electronics and integration of safety systems is bringing the vision of accident-free driving closer to reality.

3.2.1 Human factors

In addition to injury mitigation, a significant number of traffic and vehicle systems introduced into production have contributed to increased road safety. Because active and passive systems have developed further, integration of both drives further positive development. Some elements of active safety are mentioned in the following paragraphs.

As they interact in traffic humans play a major role in accidents. Accident investigations indicate that the driver is responsible for almost all accidents. Driver mistakes have varied causes. The driver is influenced by his or her health and physical condition, driving qualifications and experience, orientation abilities, climate, and comfort conditions within the vehicle. Results of an accident analysis from various sources find that approximately 65–70% of accidents are caused by human based failures, 20–30% by environmental failures, and only a minor percentage by vehicle related failures. For drivers in Germany, Figure 3.6 shows the frequency of these different factors.

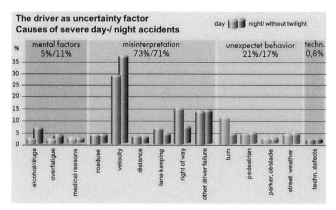

Figure 3.6: Possible misjudgment by the driver.

The mental and physical condition of a driver is one of the important factors in accident causation. A momentary over-estimation of driving capabilities or a reduced physical condition due to the consumption of alcohol or drugs affects the safe driving ability of the driver. Figure 3.7 shows the importance of the driver's

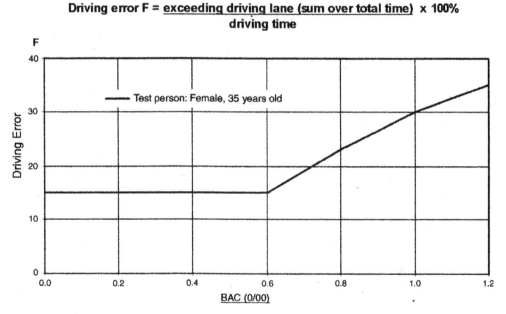

Figure 3.7: Driving failures as a function of blood alcohol content (BAC).

mental and physical condition in accident causation and demonstrates the high risk that high alcohol levels pose to drivers [3-4].

Past research and development efforts have attempted to prohibit alcohol impaired drivers from operating vehicles; however, real success in this endeavor has not been achieved.

3.2.2 Comfort and ergonomics

The vehicle can support the well-being of its driver with appropriate ergonomic systems. There is a direct relationship between driver comfort and vehicle safety. If the driver feels comfortable in the vehicle, he or she will be better equipped to react to a dangerous situation. Otherwise, because of a non-optimal driving condition or reduced attentiveness, the driver might not be able to react appropriately in a potential collision. As we will see later, a substantial number of collisions are caused by drivers who do not react in an appropriate way in a critical scenario. Thus, all features that contribute to a more comfortable drive also help to reduce accidents. These include comfortable seats, low noise and vibration levels, air conditioning, forgiving driving behavior, easy steering wheel torque, good visibility, electrically adjustable and heated mirrors, automatically adjustable interior rearview mirrors, parking support systems, and adaptive lighting systems.

Special attention has to be given to the ergonomic design of the working place "driver." The design criteria cover the 5% female up to the 95% male, which covers most of the driver population. The ergonomic layout of the vehicle controls can be checked using the computer model "RAMSIS" (Figure 3.8).

Figure 3.8: Computer simulation model "RAMSIS".

Of critical importance to accident avoidance is also the field of view via the front windshield and the rear view mirrors, currently supported by a rear view camera and with night view support in some cars. Details on these systems will be found in the following chapter. On the other hand, headlamp technology has made tremendous progress as illustrated in Figure 3.9.

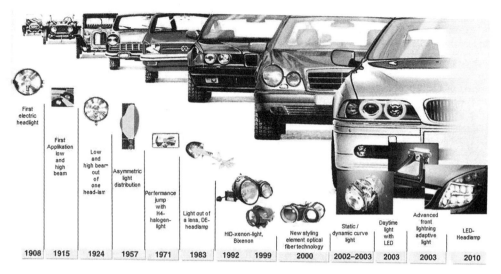

Figure 3.9: Historical development of headlamp technology [3-5].

3.2.3 Chassis and tire design

Requirements for chassis design have also increased over the last decades not only because of increased legal requirements but also because of changing customer demands. Table 3.1 [3-6] demonstrates requirements for the chassis layout, and Table 3.2 [3-7] describes the design expectations of vehicle tires.

The main requirements for chassis design include the following: integration of ABS and ESC, light weight, high quality, resistance against low speed impacts, and precise tracking.

For traction reasons we also find a high number of four wheel drives even in hybrid vehicles where the electric drive could be used for the four wheel traction. An interesting aspect of safety improvement is the increasing number of vehicles where the brakes are automatically applied or, in a preliminary stage, the driver gets a warning signal to apply the brakes in emergency situations. Although the tire looks like a minor vehicle component to most drivers, Table 3.2 illustrates the demanding design criteria of this vehicle component. A big challenge is the optimization between the handling characteristics and the rolling resistance of tires. Today reduction of the noise level at the tire-road surface has a high priority, especially in Europe.

Reducing the consumption of fossil fuel to achieve low CO_2 levels is one of the most stringent requirements for vehicle development. In addition to the reduction of weight and the optimization of the drive line, a low aerodynamic drag value is important. In the optimization process of the vehicle, it is important that the lift coefficient at the rear axle does not increase to an unsafe level while designers focus on achieving a low drag coefficient (CD) value. Figure 3.11 demonstrates this challenge.

31

Table 3.1: Assessment criteria of the chassis related to dynamics, comfort and safety [3-6]		
Subject	Assessment criteria	
driving dynamics	• vehicle handling • straight-running properties • steering and brake behavior • break away prevention • bump steer • body control	• initial steering behavior • directional precision • slalom stability • dry traction • self-steering behavior • load-alteration effect
vehicle comfort	• body acceleration • roll suspension • roll gradient • squat and brake dive • wheel dampening • edge sensitivity • bounce • copy behavior	• absorption capacity • vehicle oscillation • steering / brakebehavior • body shake/oscillation • steering shake/oscillation • steering jumpiness
driving safety	• curve control • high speed control • control during difficult road conditions • smooth road control • smooth response in event of driver error	• predictive capability • street feel • brake distance • brake precision • wheel slip control • driver assistance

Table 3.2: Expectations of vehicle tires [3-7]		
Riding comfort	Steering performance	Driving stability
suspension comfort	steering precision in 0° range	driving straight stability
noise comfort	steering precision in proportional range	curve stability
running smoothness	steering precision in stability limit	braking in curves
Road Adhesion	Durability	Economy/Environment
traction	structural durability	life expectancy
brake distance	high speed proficiency	rolling resistance
lap time	burst pressure	retreadability
aquaplaning	puncture resistance	passing noise

Figure 3.10: Chassis of the Volkswagen Golf VI [3-8].

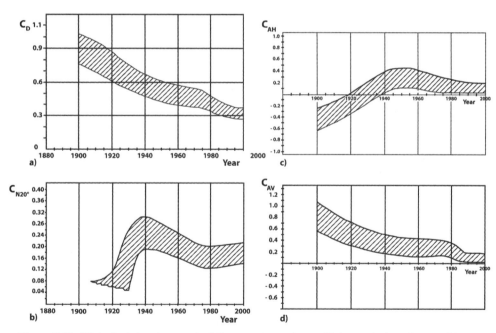

Figure 3.11: Historical development of the aerodynamic coefficients a) air resistance b) yaw moment c) lift on the rear axle d) lift on the front axle [3-9].

3.2.4 Vehicle assistance systems

ABS prevents the wheels from locking during full-on braking or on a slippery road surface, thus maintaining the steering capabilities of the vehicle. Locked wheels cannot transmit any lateral guidance forces, resulting in an uncontrollable vehicle. To prevent this from happening, the ABS control unit uses wheel rotation sensors to measure the rotation speeds of all of the vehicle's wheels. If one wheel is threatening to lock, then a solenoid valve in the control unit of the anti-lock brake system reduces the brake pressure of the corresponding wheel until it once again turns freely. Following this, the pressure is increased back up to the locking threshold. The vehicle remains stable and steerable. The driver senses that the anti-lock braking system is operating by a slight pulsating sensation at the brake pedal. Within the control range of the anti-lock brake system, the vehicle can still be steered without difficulty in spite of maximum deceleration. Consequently, ABS makes it possible for drivers to avoid obstacles and prevent collisions.

The brake assistant (BAS) helps the driver during emergency or panic braking (Figure 3.12). Based on the actuation speed of the brake pedal, it detects whether the driver wishes to perform full-on braking and, as long as the driver continues to keep the brake pedal fully pressed, increases the brake pressure automatically until the ABS control range is reached. When the driver reduces the brake pressure, the system reduces the brake pressure back to the specified value. The brake assistant can shorten the stopping distance significantly, and the driver scarcely notices the operation of the system.

ESC detects critical driving situations, like skidding potential, and takes specific measures to prevent the vehicle from spinning around its vertical axis. To allow ESC to respond to such situations, the system requires a constant supply of information, for example, the direction in which the driver is steering and the direction in which the vehicle is traveling. The system obtains driver steering direction from the steering angle sensor and the wheel speed sensors. Based on this information, the control unit calculates the nominal steering direction and a nominal driving behavior of the vehicle (Figure 3.13). Additional important data include the yaw rate and the transverse acceleration of the vehicle. Using this information, the control unit calculates the actual status of the vehicle. ESC prevents the vehicle from becoming unstable when cornering, which can occur if the speed is not adjusted

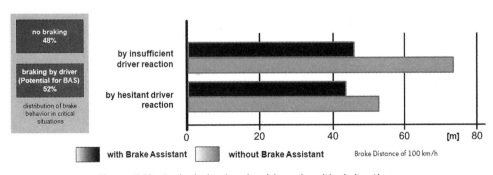

Figure 3.12: Brake behaviors by drivers in critical situations.

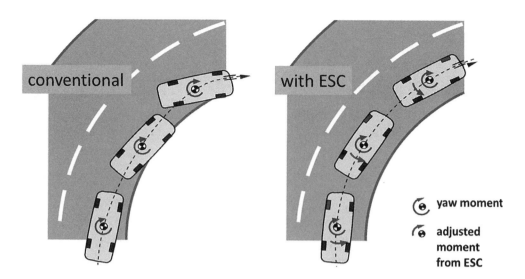

Figure 3.13: Electronic Stability Control (ESC).

correctly, if there is an unforeseeable change in the surface or condition of the roadway, or if an avoidance maneuver is suddenly required. It does not matter in this case whether the instability is expressed by under-steering (the vehicle slides towards the outside edge of the curve even though the wheels are turned inwards) or over-steering (the end of the vehicle swings out). The computation unit of the electronic stabilization program detects the type of instability using the data provided by the sensors and controls the correction by intervening in the brake system and engine management. In case of under-steering, ESC decelerates the rear wheel on the inside of the curve. At the same time it reduces the engine power output until the car has stabilised again.

ESC prevents over-steering by specific application of the front brake on the outside of the bend, as well as intervention in the engine and gearbox management. Increasing experience and significantly more sensitive sensors are permitting development of this complex control system on a more and more sophisticated level.

In contrast to conventional headlights, adaptive light systems are now able to dynamically adapt the way they illuminate the road based on different driving conditions. Xenon headlights with dynamic and static corner light improve illumination of corners up to 90%. From a speed of 9.3 miles per hour (mph) (15 km/h) onwards, the headlight cones are pointed in the same direction as the steering wheel rotation by a maximum of 15° (Figure 3.14).

In this case, the light is always distributed according to the speed of the vehicle. In dynamic cornering light, the light cone follows the steering wheel rotation at a driving speed of 9.3 mph (15 km/h) or more, to ideally illuminate the course of the bend This allows the driver to detect the course of the road, as well as any objects on and adjacent to the roadway. Servomotors in the headlight units swivel the headlights by using information about the steering wheel angle and the driving speed.

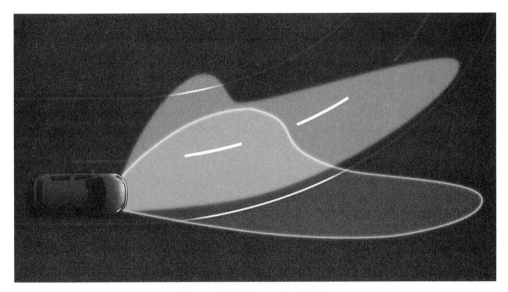

Figure 3.14: Cornering light.

The swivel angle of the dynamic cornering light is restricted to 15 degrees to avoid affecting oncoming traffic.

Another function of the adaptive front lighting system (AFS) is its ability to adjust the lighting according to speed and the road type on which the car is traveling (Figure 3.15). This means that it is possible for the headlight cones to be lowered and swivelled apart during town driving to improve close-range visibility. On secondary roads and highways the headlight cones are swivelled together again and moved further upwards to better illuminate the area in front of the vehicle. This means that the driver is able to notice and focus on objects that are farther away.

At intersections and during turning, a static turning light can significantly improve the visibility of objects, cyclists, or pedestrians. In this case, an additional

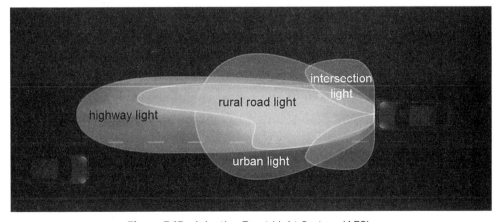

Figure 3.15: Adaptive Front Light System (AFS).

light (which expands the cone by up to 35°) is activated in the headlight when the steering wheel is turned by a sufficiently large angle or if the turn signal is activated.

3.2.5 Driver assistance systems

3.2.5.1 Longitudinal guidance systems

Currently, adaptive cruise control (ACC) is available for highway driving in newer vehicle models (Figure 3.16). In this radar-based system, the car adjusts its speed to the traffic conditions, within certain limits, without the driver having to adjust the speed manually. Once a desired speed and distance to another vehicle have been selected, the vehicle automatically adjusts itself to these parameters. However, this only happens if the other vehicle is traveling at less than the desired set speed of the vehicle with its ACC activated. Relevant objects are detected by a radar system and course of the vehicle which is obtained from dynamic data.

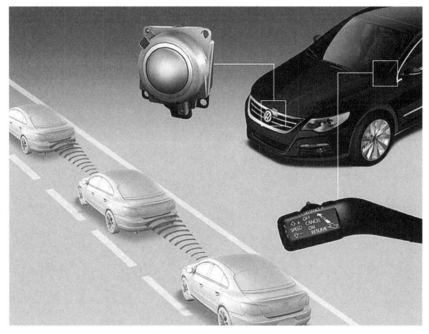

Figure 3.16: ACC [3-10].

Depending on the version of the ACC that is operating, there are various radar detection modes for following traffic as well as automatic adaptation of the acceleration behavior based on the surrounding traffic as detected by the radar system. The ACC convenience function has been supplemented by elements of vehicle safety.

If the environmental monitoring system detects a critical approach situation, the vehicle front assist prefills the braking system and increases the sensitivity of the hydraulic brake assistant to prepare for anticipated emergency braking. Both of these measures increase the effectiveness of the driver's braking. In addition, a warning is triggered, which may consist of a visual and audible warning and a brake jolt. In situations with reliable detection, automatic emergency braking is initiated when the driver does not intervene. The vehicle response is intended to provide active accident avoidance depending on the situation or reduce the consequences of a collision.

3.2.5.2 Lateral guidance systems

Side assist, as illustrated in Figure 3.17, supports the driver by alerting him/her to approaching vehicles before or during a lane change. An intuitively comprehensible feedback using warning lamps in the outside mirrors informs or warns the driver, within the system limits, about vehicles within an area behind his/her own car and in the blind spot. This makes it easier to estimate driving situations and avoid danger. The system is configured so it only warns the driver in relevant situations; there is no warning for stationary objects or oncoming vehicles, for example. Also, the system ignores vehicles that are more than one lane away to avoid confusing the driver with unnecessary warnings. The system is activated at a speed of 37.2 mph (60 km/h) meaning that it is not active in dense urban traffic. The main objective of side assist is to provide additional convenience to the driver. However, it cannot and should not relieve the driver of his/her responsibility to drive carefully and look in the rear-view mirror when passing, merging, or changing lanes.

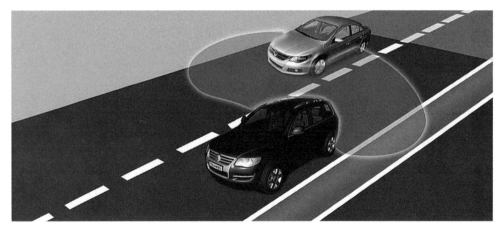

Figure 3.17: Side assist [3-4].

Another system for increasing safety is lane assist (Figure 3.18) which can be activated at the driver's request. A driver's concentration can waiver, particularly on long and monotonous journeys, allowing the car to drift off the road. This can be avoided by a slight correcting intervention in the steering based on detection of the lane markings by camera sensors. If the markings can be detected and the vehicle's speed is above 40.3 mph (65 km/h), then the system activates and indicates its status to the driver via the instrument panel. If the car moves outside of its lane, then the lane assist applies a counter-steer effect. The driver should not take his or her hands off the steering wheel when the system is active, since the system interprets this as a loss of driver attentiveness in which case, the vehicle triggers a take-over request to the driver and switches itself off.

Figure 3.18: Lane assist [3-11].

An additional lateral guidance system available at low velocity is the park assist. The park assist helps to reduce accidents during parking situations. It detects the size of a parking space by using ultrasonic-based sensing systems on the front and rear of the vehicle. Park assist calculates the optimal steering needed to position the vehicle for parallel parking and automatically turns the steering wheel so that the vehicle drives backwards into the parking space (Figure 3.19).

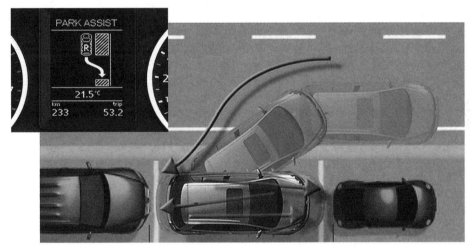

Figure 3.19: Park assistance [3-12].

3.2.5.3 Night assistance

Darkness increases the risk of accidents for drivers and pedestrians. Lower light intensity, lower contrasts, and reduced visibility make driving at night more difficult. Therefore, illuminating the environment as much as possible is important to increase safety [3-13]. The difference in effective illumination is illustrated by the difference in low beam and high beam headlights as illustrated in Figure 3.20. Although the difference in the two images is considerable, only about 25% of all drivers use high beam headlights when the traffic situation allows their use.

Figure 3.20: Example for low and high beam headlights.

To increase visual perception for the driver, more and more assistive light functions have been offered recently. In night assistance systems, Figure 3.21, there is a difference between night vision systems that show the area in front of the vehicle on a display screen and high beam assistance systems that adjust the light distribution maximizing driver visibility while at the same time minimizing glare for other vehicle drivers.

Figure 3.21: Night vision assistant with pedestrian marking [3-4].

Night vision systems record the area in front of the vehicle using a camera. Either passive or active systems are used. Passive systems use a thermal imaging camera that is sensitive to all heat sources in its field of view (e.g., pedestrians, vehicles) and clearly illuminates them on a display in the vehicle. In addition, the system can use intelligent image processing to classify pedestrians and mark them on the display, if required. This is helpful, for example, when a pedestrian is crossing the road.

When night vision assistance systems are activated, the area in front of the vehicle is illuminated by an infrared (IR) light with a range of approximately 492 feet (150 m). The IR light reflected back from the surroundings is detected by a normal vehicle camera that is also sensitive in the infrared (IR) range. In contrast to the passive system, the driver is shown a completely illuminated picture of the surrounding area on the display.

High beam assistance systems can also detect relevant objects at an early stage. These use a camera to observe oncoming traffic and vehicles in front a driver's vehicle, and adapt the light distribution of the headlights accordingly. One variant is the high beam assistant. The high-beam assistant permits the high beams to be switched on and off automatically depending on the traffic situation. The high beam is switched off if sensors detect oncoming traffic or vehicles in front of a driver's car. Once the area in front is free again, the high beam is automatically reactivated [3-14]. This gives the driver optimum visibility without affecting oncoming traffic.

The high beam assistant is activated during one of several scenarios: when it detects a vehicle approximately 1312 feet (400 meters [m]) in front of it with the vehicle's exterior lights switched on; when it detects an oncoming vehicle up to 3280.8 feet (1000 m) away with its exterior lights switched on; or when street lights are on in response to an ambient brightness level of less than 0.5 lumens (lm) and when the vehicle speed is above 37.2 mph (60 km/h). If one of these conditions is no longer applicable, then the high beams are automatically switched off.

In driver assistance systems, the "eye" of the system is the front camera, which is installed in the base of the rear-view mirror. The control unit electronics are installed in the rear-view mirror housing. The control unit is connected to the data bus network via the CAN driveline data bus. The camera detects light signals in the visible range up to a maximum of 3280.8 feet (1000 m) in front of the vehicle and an opening aperture of 30° in the travel direction (Figure 3.22).

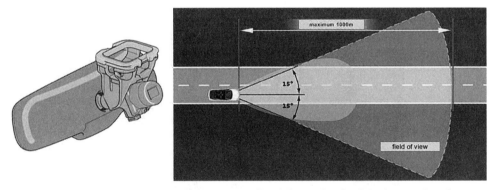

Figure 3.22: Camera which provides surrounding information for functions such as lane departure warning or high beam assistant.

A further development of the high-beam assistant is the adaptive cut-off headlight (Figure 3.23) which adapts the light range automatically, so the area in front of the car is optimally illuminated and the driver's field of view extended. A camera not only establishes that there are other vehicles in front, but also identifies their positions. This information makes it possible to adapt the light and dark boundary (LDB) of the headlights smoothly between low and high-beam distribution, depending on the range of the oncoming vehicles or vehicles in front of the camera.

An additional development is the anti-glare high beam (Dynamic Light Assist–Figure 3.24). In this case, light distribution is not just adapted vertically, but can also be swivelled horizontally. This makes it possible to mask out the other road users in the high beam, thereby maximizing the driver's visibility, since it is only switched off when it would affect surrounding traffic.

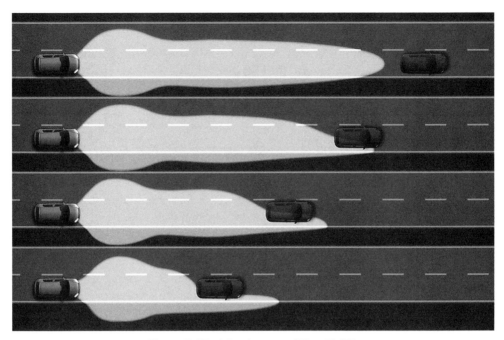

Figure 3.23: Adaptive cut-off line [3-15].

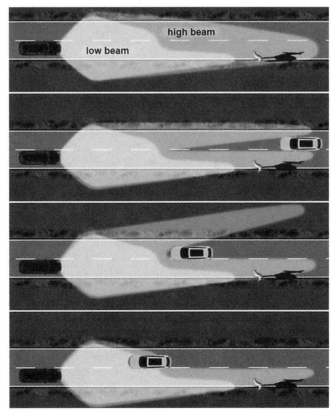

Figure 3.24: Anti-glare high beam (dynamic light assist) [3-11].

Compared to the adaptive cut-off line the anti-glare high beam not only adapts the headlight range, it also dims selected areas where other road users could be blinded. It also improves the visibility in areas next to the road for oncoming or preceding traffic. Accident reports show that such systems have a high potential to reduce the number of accidents because they enhance visibility and perception for the driver. A comparison between halogen and xenon headlights is illustrated in Figure 3.25. Improved illumination systems could reduce many driving and loss-of-control accidents because they are often a result of an incorrect perception of the vehicle's speed as a result of low illumination [3-13].

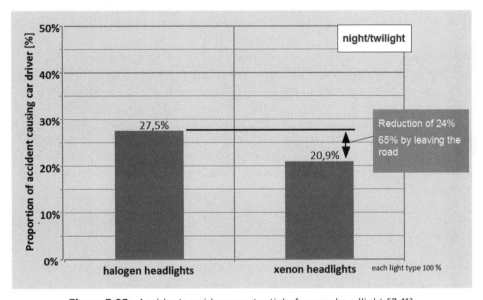

Figure 3.25: Accident avoidance potential of xenon headlight [3-11].

In the future, information from surrounding vehicle sensors will make it possible to expand the existing functions of illumination systems significantly. Running lights may be configured with a predictive function, in contrast to their current reactive functions. Using a camera and highly accurate navigation data, it will be possible to determine the precise course of the road ahead of a vehicle. Using this information, for example, a cornering light can be swivelled into a corner before the steering wheel is turned. In this way, the driver will be able to see the course of the road and potential obstacles even earlier than is possible now. In urban driving, intersections may be illuminated in a predictive way, there by, allowing crossing traffic to be detected before the vehicle enters the intersection.

The light emitting diode (LED) is the light source of the future in automobiles (Figure 3.26). Greater efficiency, higher light intensities, and declining prices are making this technology more and more attractive for improved design and CO_2 reduction. LED technology reduces CO_2 because it is much more efficient than incandescent technology and consumes less energy [3-16]. In addition to its environmental advantages, it increases the recognition level of cars by introducing new, individual light designs and providing new light assistance and safety functions.

Figure 3.26: LED daytime running light [3-17].

Today LEDs are used as signal lamps in many vehicles because their low energy consumption and long service life represent clear benefits compared to incandescent bulbs. Daytime running lights are a special front light designed to increase the recognition level of the vehicle in the daytime. While their intensity is lower than low-beam headlights, their operating period is much longer. To save energy and provide a more flexible design, almost all carmakers use LED technology for daytime running lights. As a result, daytime running lights not only increase safety, but provide the automobile with a distinctive design element that provides a high level of model or brand recognition.

Because of their short switching times and compact design LEDs can also support conventional light assist functions such as cornering lights and masked continuous high beams. LED technology enables dynamic modification of light distribution depending on different traffic situations. Most of the lighting applications described need a variable light actuator. One way to illuminate the different sectors in front of the car is to use a LED matrix beam. The LED matrix beam is an array of high brightness LEDs in a headlight.Figure 3.27 shows the configuration used by the masked high beam.

For the masked high beam a camera observes the environment in front of the car, then a controller calculates the distance and the lateral position of cars in the scene

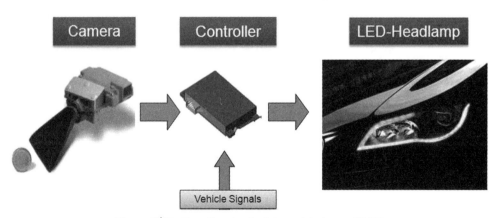

Figure 3.27: Functional principle matrix-beam [3-18].

Figure 3.28 / 3.29: Examples of marking light (right) with a LED matrix beam (left).

from the camera information. Finally the matrix beam dims or switches off several LEDs so that other road users are not blinded.

An LED matrix beam makes it possible to illuminate or mask specific areas in front of the automobile, in order to increase the driver's ability to recognise objects and the course of the road (Figure 3.28/3.29). Surrounding sensors detect objects in front of the vehicle and evaluate their potential danger. Following this evaluation, dangerous objects are displayed for the driver using a marking light to attract the driver's attention and reduce his or her reaction time. Thus, light-based assistance systems try to optimize the perception of the driver by directing as much light as possible on critical objects while reducing glare to address higher accident risks at night [3-19]. For example, using a marking light, a driver can detect a pedestrian who is walking on the road earlier than he or she can without the marking light. The marking light redirects the driver's attention to the pedestrian. Objects are illuminated by light spots after they are detected by environmental sensors. This is one example of a direct light-based warning system that could substitute for displayed warnings that take additional time for drivers to process.

Another active safety light is an advanced light functionality that illuminates the potential escape path for a driver at risk of an imminent accident. After the target car crosses the trajectory of the driver's vehicle, the active safety light shrinks the headlights of the driver's car so that it only illuminates its own lane and shines strongly at the tail of the target car. Figure 3.30 illustrates this function. The light

Figure 3.30: Escape light function as a direct light-based warning that could substitute for warnings on displays.

is constantly adjusted while the oncoming car is crossing the road in front of the driver's vehicle. With this light function the driver is able to steer his or her car in an evasion maneuver instead of into the path of the oncoming vehicle and avoid a collision.

3.3 Driver, vehicle, and environment

3.3.1 Introduction

In order to detect critical driving situations, a range of different information is required from the vehicle and the environment. Internal vehicle information such as speed, or dynamic driving information such as transverse acceleration and ESP and ABS activation, can be obtained directly from the bus system (CAN, Flexray). Surround sensors integrated into the vehicle can provide information about the surrounding environment. Other moving objects as well as static obstacles can be detected in this manner and the areas to the sides of the vehicle can be monitored.

Furthermore, the configuration of the human-machine interface (HMI) of the system must take increasing account of the driver. In particular, the functional design between the system and driver must be configured so that the driver intuitively understands the function and operation of the system. Driver modeling (i.e., performance, driver type, risk acceptance level, current mood) can be used for adjusting the system interface with the driver. As a result, the driver will accept the assistance system, increasing its value.

Each of the aforementioned areas represents different possibilities for modeling the driver. Performance describes the current ability of the driver to drive a vehicle. Low performance ability means that the driver cannot detect dangerous situations or cannot react to them appropriately because of fatigue or distraction. Prediction of driving maneuvers provides information about the driver's current mood. If the driving maneuver currently being undertaken is not performed in the familiar way, this can allow conclusions to be drawn regarding stress or aggressive driving behavior. The driver type describes the limits within which the driver feels safe and has his or her vehicle and the driving situation under control. Risk acceptance level describes the level to which the driver deviates from a defensive driving behavior in his or her driving. These different models of the driver can improve various aspects of assistance and safety systems [3-20, 3-21].

System intervention is based on the determined hazard level of the current driving situation. Normally, this is obtained from the status of the driver's own vehicle and that of other road users, information which can be obtained using sensors. In order to achieve sufficiently adequate perception of the periphery and interpreting it with the driver's own vehicle, it is essential to precisely establish various parameters while driving. For this reason, intelligent sensors send their data via the bus system cyclically. In particular, measurement of signals relevant to the chassis is required in many sub-functions. For example, the speed, yaw rate, and pitching and rolling movements are measured with high resolution and update rates. Both the components of the environmental perception system and modules which predict critical situations require these data to model the situation which is required for predictive safety functions. However, additional information can be used such as data from digital maps and knowledge from driver modeling. The more complex a situation is, the more information is required to ensure safe intervention.

3.3.2 Driver modeling

The goal of driver modeling is to build a model of the driver and his behavior in the context of driving a vehicle. The first driver models were mathematical control models that incorporated human limitations, such as a limited field of view or errors in vehicle control such as steering or accelerating. These models provided a more realistic controller that could be used to, for instance, fine-tune the vehicle's behavior in a simulated race track. There are many other aspects of the driver that have been investigated in recent years, such as models that detect the driver's emotional state, distraction, fatigue, driving behavior or driver type (sporty, normal, defensive), and risk acceptance level. In general, modeling of driving behavior can be divided into strategic, tactical, and control layers, depending on the duration of the task. In the context of driver assistance and safety systems, predicting driving behavior on the tactical level, the level of individual maneuver, is of special interest. If the driver's intention is known, that information can be used to adapt the system accordingly. One example is the behavior-dependent adaptation of longitudinal vehicle control [3-20]. These adaptations can increase the acceptance level of the driver assistance and safety system. Another example is the fatigue detection systems currently available in a variety of commercial vehicles that suggest a break when the driver shows signs of fatigue.

Driver modeling can also provide valuable information to adapt active safety systems (Figure 3.31). Prediction of driving maneuvers provides information about the driver's current mood. If the driving maneuver currently being undertaken is not performed in the familiar way, this can allow conclusions to be drawn regarding

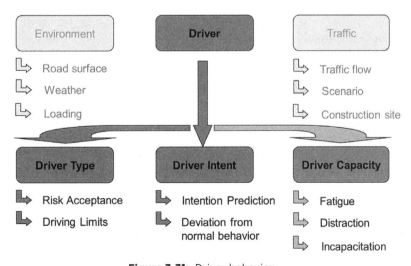

Figure 3.31: Driver behavior.

the driver's state or aggressive driving behavior. The driver type describes the limits within which the driver feels safe and has his or her vehicle and the driving situation under control. Risk acceptance level refers to the extent to which the driver endangers himself or herself and the surroundings by his or her driving behavior. Sporty drivers certainly tend to have a higher risk acceptance level, but this is not necessarily the case. These different models of the driver can improve different aspects of assistance and safety systems.

Many current active safety systems, especially autonomous braking systems, assume that the driver will intervene to avoid a collision and are only activated when a collision is unavoidable. However, investigations of accident databases have shown that drivers often do not react in dangerous situations. The results of such an investigation are presented in Table 3.3 [3-22].

Table 3.3: Actions of the driver in rear-end crashes [3-22]			
	Approach to braking vehicle	Approach to stationary vehicle	Approach to vehicle driving at constant speed
No action	81.4 %	78.4 %	83.8 %
Braking	12.2 %	15.5 %	8.1 %
Steering	1.1 %	2.2 %	1.7 %

The driver's failure to act can have many causes, such as distraction or misjudgement. A system that predicts the intention of the driver might be able to recognise a situation of this kind and include these data in its judgement of the situation. This means that valuable time is lost when a system assumes that the driver can still react, even if the driver is unable to do so. A system that detects the driver's failure to act to the situation can potentially increase its effectiveness by early intervention.

Even those drivers who do react in dangerous situations often do not trust themselves to brake strongly or to make sudden steering corrections. An investigation of last possible avoidance maneuvers has shown that drivers differ markedly in their judgement of dangerous situations. Reference [3-23] shows how threatened candidates felt during different risky driving maneuvers. It is apparent that many candidates categorised the situations with a time to collision (TTC) of 1.5 to 2 seconds as dangerous, although these situations were significantly beyond the unavoidability threshold; it would be impossible to avoid a collision using physical driving maneuvers. The individual differences shown in [3-23] do overlap with some investigations. As part of a candidate study, drivers were instructed to drive towards a stationary object and then avoid it at the last possible moment. Figure 3.32 shows the TTCs for the tests, at which point an avoidance maneuver was triggered. The values for the TTC are in the range from 0.8 seconds (s) to 1.2 s before contact.

The potential of driver-adaptive systems can be seen in these figures. In the case of drivers who did not trust themselves to make any risky maneuvers, this knowledge can be used to predict driver-specific unavoidability, which occurs significantly before the physical driving unavoidability on a temporal scale.

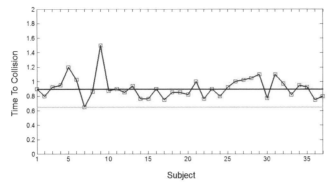

Figure 3.32: Investigations of the last possible avoidance maneuvers [3-24].

The potential improvement in effectiveness of an autonomous braking system based on the prediction of driving intent was studied by Bauer [3-24]. In a simulator experiment, 20 drivers drove for approximately 30 minutes in an inner-city environment. With that driving data, driver models predicting the drivers' stopping behavior were generated for each individual driver. At the end of the simulator experiment, an unexpected strong brake activation of the leading vehicle produced a rear-end collision. An algorithm detected if the driver was reacting appropriately based on his or her previous stopping maneuvers, and the system performed a full brake in those cases were no driver reaction was observed and the situation criticality exceeded a driver-dependent threshold. Of the 20 drivers, 5 did cause a collision at the end of the journey, and some even collided with a preceding vehicle during regular driving. Table 3.4 shows the TTC values at which the full brake would have been performed in all of these cases [3-24].

Table 3.4: TTC for system intervention [3-24]		
Driver ID	Collision 1	Collision 2
4	1.46 s	—
10	1.33 s	1.3 s
12	1.14 s	—
16	1.38 s	1.41 s
20	1.49 s	1.49 s

The experiment showed that a driver adaptive system would have been activated very early, substantially reducing collision velocity or completely avoiding the collision. In addition, no incorrect activation of the autonomous brake was observed, which could have led to a dangerous situation like a rear-end collision.

3.3.3 Vehicle data and perception

Precise determination of various parameters while driving is essential for a data collection system to adequately and accurately perceive the surrounding area and to interpret this with the system's own vehicle. For this reason, intelligent sensors send their data via the bus system cyclically. In particular, measurement of relevant chassis signals is required in many sub-functions. For example, the speed, yaw rate, and pitching and rolling movements are measured with high resolution and update rates. Both the components of the environment perception system and the modules that predict critical situations require these data to create an image of the situation that is necessary for predictive safety functions.

Information about the outside temperature and the amount of precipitation and light are available, provided by systems that focus on driver comfort; these systems also provide information about ambient conditions. This information in turn makes it possible to reach conclusions about visibility conditions or general conditions regarding the coefficient of friction, for example. Currently, approaches are being developed to integrate a tactile sense into the skin of the vehicle for the (un)certainty which is more of a feeling by the customer, so that new customer-relevant safety applications can be provided even when the vehicle is parked. Navigation data, which is increasingly being collected, can also be used to increase safety. For example, route guidance mechanisms could have their parameters set differently in urban areas than on a highly developed highway.

3.3.3.1 Crash prediction

Today's automotive crash- and pre-crash systems provide drivers with high-level safety-related features beyond the requirements defined by laws. Safety systems with irreversible actuators like those used in airbags require the engineer to investigate different parameters like functional safety, customer satisfaction, and product liability. System intervention is based on determining the hazard level of the current driving situation. This is usually obtained from the status of the vehicle and that of other road users which is registered by sensors. However, additional information such as from driver modeling or map data can also be used. The more complex a situation is, the more information is needed for optimal intervention. Current safety systems do not have all relevant information. Consequently intervention normally takes place only a few hundred milliseconds before a predicted collision. Only in this period can a collision no longer be avoided by the driver, no matter what the driving situation. Intervention can potentially take place sooner with increasing knowledge about the situation, without impairing safety.

The challenge in applying time-critical safety functions is having an adequately large detection rate of critical scenarios where activation is desired, while ensuring that inaccurate triggering is kept at a minimum. This is necessary to make sure that such safety functions will be accepted by customers. The topic of product liability is also critical, since inaccurate triggering, particularly in non-reversible systems could result in unacceptable outcomes or costs. In addition to ensuring that the system's implementation decision is correct, it is also necessary to guarantee the reliability of the function. Therefore, it is necessary to quantify the detection and false-alarm rate equally on the basis of representative measurement data and on accurate extrapolation based on these data.

A single sensor decides whether or not the car will collide with an obstacle. This decision is made by evaluating the data from the corresponding sensor together with the data from the car's other sensors. First of all, the system evaluates whether the car can avoid the obstacle by a strong steering maneuver. After this evaluation, the quality of the data is checked to ensure that the decision is based on accurate and reliable data. If the decision is based on questionable data, no alarm is given to the driver.

To process the data, a binary decision tree can be used [3-25]. According to the requirements mentioned above, this binary decision tree can divided into three major parts as seen in Figure 3.33. In the main part of the decision tree, the pre-crash detection, the system evaluates whether or not a crash with an obstacle can be avoided. The quality of the track data is reviewed in the track validation box. If the data are accurate, they are passed along to the next step. Within the situation analysis the current situation is examined. In some situations no alarm will be given, depending on the actuator used.

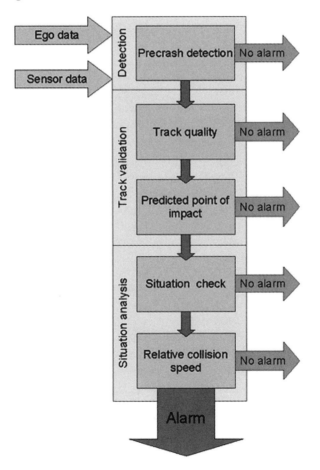

Figure 3.33: Binary decision tree [3-25].

The data traverses the binary decision tree from top to bottom, passing each of the modules of the tree. In each module the data are verified to fulfil a specific requirement. Only if all modules are passed, is an alarm given for the single sensor.

Every sensor type provides a periodic digital measurement or snapshot of the real situation. The accuracy predominantly depends on frame rate and signal-to-noise ratio. The parallel calculation of different sensors will provide different results for predicted crash times. Preparing a more reliable crash decision is the aim of the combination of sensors. Because it is difficult to evaluate a complete system when the logic part is not validated, the functional requirements could be combined in a crash detection unit, so that it is possible to quantify requirements for each sensor.

To have a manageable number of possible situations, only the driving state and driving scenario could be taken into account as inputs for the arbitration unit that makes the decision. However, driving is complex. According to their properties and performance, especially in high dynamic situations, sensors will either detect the crash or not. One simple way to combine single decisions is to deploy if at least two of three sensors detected the crash. Accident research and analysis show how often accident situations occur worldwide. It is obvious to abstract these for sensor testing on proving grounds. Test-drives show the performance of the system using combination of single crash decisions. Furthermore potential for improvement of sensors and combinatorial logic might be identified.

3.3.3.2 Evaluation

The evaluation of each sensor and different sensor sets has to consider two major aspects. Objects inside the predicted path of the vehicle, that sensors do not detect or classify as non-hazardous may lead to false negatives. At the same time sensors that report non-existing objects may cause false positives. For this reason, the design engineer has to take into account both of these situations. Consequently, major attention has to be paid to false positives, which have to be reduced to a very low level. Fulfilling this requirement always leads to a trade-off with the detection rate of crashes. Evaluation during test drives has to prove that this trade-off represents the optimum between false positives and negatives.

Unfortunately it is not known a priori, in which situations a false positive alarm may occur. So evaluating this situation has to be done by recording and analyzing false positives during extensive road tests. Since it is obviously not feasible to test drive billions of kilometres to reach extremely low failure rates, low failure rates have to be proven not by road tests but by statistical extrapolation. This is only suitable if the chosen scenarios (i.e., area, road types, weather conditions) are sufficiently representative to allow statistically defensible data.

The decision that a crash is unavoidable is made using Boolean AND gating of multiple attributes derived from sensor data. Each attribute is compared to a chosen threshold as shown in the binary decision tree in Figure 3.33. If the actual value exceeds the threshold of a gate, the decision process continues with the next gate in the tree until a gate blocks or the last gate is passed which means that an unavoidable crash is detected. The overall probability, P_{fa}, of a false alarm in the procedure described above may be calculated by simply multiplying all single probabilities of each gate, P_{gi}. This probability can be derived from the attribute's probability density graph that was recorded during test drives. Figure 3.34 [3-26] shows a graph for a measured attribute, where two different hypothesizes (H_0, H_1) are

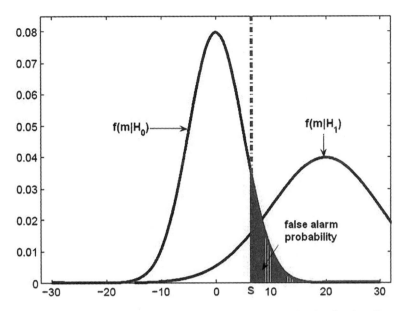

Figure 3.34: Exemplary description of probability densities on basis of a signal's attributes.

distinguished. In this case these hypotheses represent two cases H_0 (crash avoidable) and H_1 (crash unavoidable). For a given threshold S, the false alarm probability can be obtained by integrating the graph f (m|H0) from S to ×.

Otherwise, if the probability density is known, it is possible to calculate the threshold needed to achieve a given rate of false positives. In practice, the challenge is to record all graphs of probability density for each chosen attribute, where it has to be assured, that a sufficient and representative amount of road test data has been taken into account. A necessary indication of a sufficient amount of measurement data may be, in this case, the convergence in probability density of all sensor parameters itself. If all probability graphs are known, the next challenge is to find a suitable set of thresholds that achieves, in combination (AND-gating), the demanded low rate of false positives while keeping the detection rate constantly high [3-26].

3.3.3.3 Environment detection
Cruise control did already exist in the U.S. in the beginning of the 1960s, based on a mechanical solution. Since Mercedes-Benz fitted its first S-Class with a electronic cruise control system in 1998, predictive sensor technologies have undergone rapid development. Using measuring systems based on radar, lasers, and cameras, various assistance and safety functions now make driving more convenient and safer. During the initial years, radar systems tended to be preferred; however, it now appears that other measurement technologies offer advantages for more complex environmental perception. The requirements for integrating vehicle safety functions are increasing while the functionality of predictive sensors and representation of the surrounding areas are not. Simple point targets were sufficient for distance control with ACC, while the trend now is toward objects with real dimensions in

three dimensional (3D) space as well as descriptions of areas that can be driven in a complex map of the surroundings.

Currently different sensor technologies are used for driver assistance and for safety systems (Figure 3.35). Some of these sensors are already available in serial cars, for example, 77 Giga Hertz (GHz) radars for ACC-functionality. Many others are in a development stage and others are in the research phase. Sensor technologies can be grouped into three categories: radar, laser, and camera technologies.

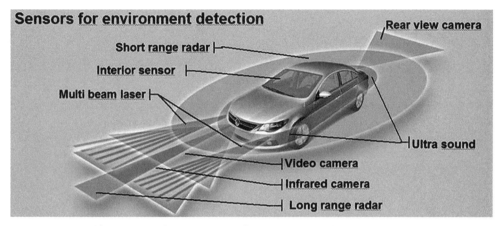

Figure 3.35: Sensor systems for environment detection [3-3].

Radar is an abbreviation for "radio detection and ranging." The sensors emit electromagnetic high-frequency waves that spread through the observation area. The waves are reflected by certain objects and deflected back to the receiving antenna. By analyzing the signal differences between transmitted signal and received signal, three characteristic parameters of the reflecting objects can be measured: range, relative speed, and angle. Using the information from successive measurements and stabilizing the sensor data by tracking provides robust object position and dynamic description of the surroundings. Radar technologies can be distinguished by their measurement principle. Pulse radars measure objects using the basic pulse echo principle. After a short transmit period, a long receiving period is required. Other radar technologies measure continuously; this means that they are constantly sending and receiving data. This continuous wave (CW) principle is more complex to compute, but offers more robust measurements; and therefore, is often used successfully. Different frequencies are currently used for automotive radar applications.

The 24 GHz ISM-conform radar (narrow band) offers quite precise localization and dynamic measurements. It was investigated for many different applications including a front pre-crash system and a side assist system. A 24 GHz UWB radar (ultra wide band) technology operates with very short pulses, some of which are measured in picoseconds. This corresponds to a very fine resolution in the range of approximately 1.97 inches (in) (5 centimeters [cm]). Until 2013, a limited frequency approval is available for maximum 7% penetration rate of the fleet (the "package solution") [3-27]. Many ACC radars operate in the 77 GHz domain. In addition, there are two measurement technologies on the market in the 24 GHz domain: pulse radar technology and CW radar technology. Both techniques offer good performance. For future applications there is an opportunity to operate automotive radar in the 79 GHz domain. While this is an interesting approach for the future, the economic implications must be investigated. Radar measurements are quite independent of weather conditions.

Typically laser sensors run in the infrared domain and measure the time-of-flight of a transmitted light impulse that corresponds with the distance to the target. At the moment two different types of lasers are used for automotive applications. In the scanning laser, the infrared impulse is generated by a laser diode and sent in different directions by a rotating mirror. The reflected impulse is received by a photodiode. Analysis of the received energy in different ranges provides the range measurement. The multi-beam laser has many transmitting elements and receiving elements that measure the surrounding area simultaneously using all of the infrared beams. Laser technology offers the possibility to measure distances and angles; however, like all optical measurement technologies it is not completely independent of weather conditions.

Imaging is a wide field of technology and includes several approaches that are often used in the automotive field. For example, mono camera technology, which involves a single camera, is often used to detect lane markings and to recognize vehicles. This technology can be used for vehicle-to-lane association, an important task for ACC. However, mono camera image processing has difficulty with object recognition and classification, although research using it for pedestrian classification has been carried out for decades [3-28]. Stereo vision, using two cameras whose optical axes are oriented in parallel with a base line, offers more robust image information than a single mono camera. Due to geometrical relationships, objects can be detected in the scene relatively easily without using demanding classification methods. If, following detection, an object class (i.e., vehicle, pedestrian, etc.) is required for a certain application (e.g., pedestrian safety measures like active hoods) a classification algorithm can be added. Some research activities show very promising approaches in that field [3-29].

Based on stereoscopic vision, which is a human property, different manufacturers are already offering stereo cameras integrated into vehicles windshields that make it possible to measure extended objects with their movement and to react accordingly (Figure 3.36).

Far-infrared cameras image thermal information in the wavelength of around 10 micrometers (μm). This kind of camera uses different technologies including uncooled or cooled imagers or bolometers. These sensors are often used for night vision applications where they enhance the visibility of heated objects like human beings against a cold background. Near-infrared cameras are typically sensitive in

Figure 3.36: Object recognition using video camera [3-3].

the wavelength around 1 μm. Near-infrared images look like images in the visible domain that are mixed with some thermal information.

It is obvious that optical imaging is affected by weather conditions reducing the maximum detection range and causing other side effects to appear. All these sensors have advantages and some characteristic disadvantages; therefore, the more challenging the requirements from pre-crash applications, particularly false alarm rates or measurement accuracy, the more sensors using different technologies have to be combined into a multi-sensor set. Especially for non-reversible pre-crash systems it seems necessary to activate a very precise fused sensor set only, because actuators need an extremely low false alarm probability to fulfil both their functional and safety requirements. Currently different fusion techniques are evaluated for safety features. Driver assistance systems typically use Kalman filtering approaches for low-level, event-based multi-sensor systems. Alternative high-level fusion concepts probably have to be considered for pre-crash applications because of their demands in terms of the short visibility of hazardous objects before collision [3-30]. The following is provided to summarize the measurement capabilities of all these technologies at a glance (Table 3.5).

Each sensor type has individual strengths and weaknesses. In order to provide an accurate depiction of the environment, information from different sensors, each of which captures certain aspects of the situation, can be combined. Here, some of the currently available sensors are presented in more detail.

GPS sensors are capable of detecting the current location of the vehicle with an accuracy of a few meters. This information can be valuable if an assistance or safety system is designed to be active only under special circumstances, for instance only on highways. However, GPS coordinates do not always allow the vehicle to be mapped onto a lane, since a variation of only a few meters can position the vehicle in the wrong lane.

This mapping can be improved, however, by also including the electronic horizon provided by a digital map. The electronic horizon offers a wide range of data regarding the driving environment, such as the current road type, the distance to the next road segment, and the most probable routes at the next junctions. By tracking the movement of the vehicle, it is often possible to infer the current lane given information from the electronic horizon. The digital map may also provide valuable information such as construction sites or other road irregularities. Neither of these sources provide information about obstacles, which makes it necessary to also include range sensors in the analysis of the situation.

As was described above, range sensors include radars, lasers, ultrasound sensors, or cameras. Which of the sensors is the best choice for a given application depends on the requirements of that application and the capabilities of the sensor.

Table 3.5: Perception capabilities of different technologies				
	Sensor Technology		Camera	
Criterion	Radar	Lidar	Mono	Stereo
Range Measurement	•	•	o	•
Velocity Measurement	•	o	o	o
Angle Measurement	o	•	•	•
Object Size Measurement	o	•	o	•
All weather capability	•	o	o	o
Object Classification Capability	o	o	•	•

Legend: • measurement, ° limited estimation capability.

Long range radars, such as the 77GHz radar, are capable of accurately detecting objects that are more than 328 feet (100m) away from the vehicle and also measure the other vehicle's velocity, which makes them a good choice for systems like ACC. However, these sensors do not perform adequately when detecting stationary objects or objects that do not reflect the radar beams well.

Cameras do not have these disadvantages and are able to classify obstacles as pedestrians, regular vehicles, or trucks. However, they are more weather-dependent than radar and have a lower detection range. Figure 3.37 shows images of the kind of data that can be measured by radar, lasers, and two different types of camera sensors. The top left image shows data from the radar. As can be expected, metal objects such as the crash barrier on the left and the vehicle on the right lane have good reflective properties and can hence be identified by the radar. It can be seen that the radar is only able to identify point-sized objects; the crash-barrier for instance is not viewed as a single object but as many different points. The multibeam-laser depicted in the top right image is able to measure the width of an object, and is therefore, able to detect the barrier as a single object. The video camera shown on the lower left image is able to detect lane markings and widths in addition to the vehicles within a certain distance of the vehicle. The infrared camera visualizes heat signatures that can be used to identify single objects.

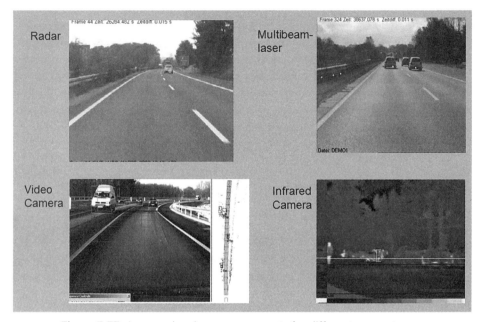

Figure 3.37: Images showing measurements for different sensor types.

Figure 3.38 depicts a fusion of several sensor sources. In this example, a GPS unit and radar provide raw data about the environment; however, both sensors are inaccurate in their measurements, as shown by the point clouds on the left side of the figure. These data lead to the model of the environment shown on the upper right side As the figure depicts, these measurement errors lead to an erroneous model showing objects positioned outside the lanes, which is of course very unlikely to be the case. The fusion of the radar objects with the lanes, as provided by a digital map,

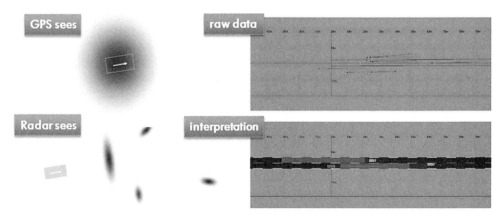

Figure 3.38: Fusion of radar and GPS data.

makes it possible to correct the inaccurate radar measurements by positioning the objects on the most likely lanes. The result of the fusion is shown on the lower right side of Figure 3.38. This is a consistent and more accurate representation of the situation than that provided by each individual sensor.

The model that results from a complex sensor fusion and situation analysis allows for the situational awareness needed by assistance and safety systems. Figure 3.39 shows the vertical integration of sensors into a fused model of the environment and the subsequent intelligent analysis of that model [3-12]. In this example, objects from range sensors and the vehicle are placed on a complex map of the environment. Given that map, the system predicts lane changes from other vehicles, such as the car merging onto the freeway on the right of the red car, and warns the driver if that lane change could result in a critical situation.

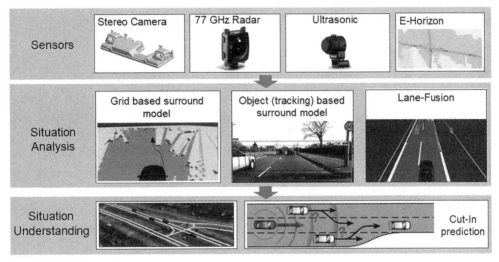

Figure 3.39: Vertical integration of sensor data into a situation aware description of the environment.

3.4 References

3-1   Gonter, M., et. al. Entwicklungstendenzen bei Sicherheitseinrichtungen moderner PKW; Symposium neue Fahrzeuggeneration, Landesfeuerwehrschule Hamburg, 2003.

3-2   Gonter, M. Moderne Rückhaltesysteme in Fusion mit der Aktiven Sicherheit, Tagung Airbags, Ludwigsburg, 2001.

3-3   Schulenberg, P., Gonter, M. Active Safety by Driver Assistance Systems; International Seminar on Automotive Electronics Technology, Society of Automotive Engineers of China (SAE-China), March 17–18, 2009, Shanghai.

3-4   Seiffert, U. Automotive Safety Handbook, 2. Edition, SAE International 2007, ISBN-978-0-7680-1798-4.

3-5   Lachmeyer, R. Handbuch Kraftfahrzeugtechnik, 6. Auflage, Vieweg+Teubner 2011, ISBN-978-3-8348-1011-3.

3-6   Heising, B. et. al. Fahrwerkhandbuch, Vieweg, Mai 2007, ISBN 978-3-8348-0105-0.

3-7   Volk, H. Reifen. Handbuch für Kraftfahrtechnik, Vieweg+Teubner, 6. Auflage, 2011, ISBN 978-3-8348-1011-3.

3-8   Wohlenberg, S., et. al. Das Fahrwerk des Golf VI. Adaptive Fahrwerksregelungundadaptive Cruise Control, ATZ,2008-6, www.ATZonline.de

3-9   Hucho, W.-H. Aerodynamik des Automobils, Vieweg, 5. Auflage, ISBN 3-28-03959-0.

3-10   http://www.volkswagen.de/vwcms/master_public/virtualmaster/de3/ unternehmen/innovation_technik/assistenzsysteme/acc.html (Oktober 2010).

3-11   Gonter, M., Grunert, C., Wegwerth, C. Sicherheitsgewinn durch aktive Lichtfunktionen Möglichkeiten und Grenzen; Optische Technologien in der Fahrzeugtechnik, VDI, Karlsruhe, Germany, April 24, 2012.

3-12   Gonter, M. Die automobile Zukunft; ÖAMTC Expertenforum, Fahrerassistenzsysteme–Gegenwart und Zukunft, Wien, November 30, 2011.

3-13   http://www.autoesthoferteam.com/Haendler/A00340/?audi&id=98000&DOM=/ haendler/modelle/a7/a7_Sportback/ausstattung/assistenz_Systeme/ nachtassistent_mit_marketing_erkannter_fussg/ (Oktober 2010).

3-14   Hella Fahrerassistenz-System: Technische Informationen.HellaKGHueck& Co., (http://www.hella.com/hella-de-de/assets/media_global/Autoindustry_ ti_fas_d.pdf (Oktober 2010))

3-15   Wegwerth, M., Gonter, M., Laschinsky, Y., Hilgenstock, J., Thomschke, S. Active Safety Light in Proceedings on Vision Conference, 2008, Paris, France.

3-16   ATZ, 10/2008, Volume 110/page 898.

3-17   Hummel, B. Die Zukunft des LED-Scheinwerfers, Audi, Haus der Technik, LED in der Lichttechnik, 10./11. March 2008, Essen/Germany.

3-18   Grunert, C., Meyer, B., Köther, G., Gonter, M., Magnor, M., Vollrath, M. Psychophysical measurement of headlight glare aftereffects on human contrast perception for optimizing a driving simulator, International Symposium on Automotive Lighting, Darmstadt, Germany, September 26–28, 2011.

3-19   Gonter, M., Bauer, C., Rojas, R. Fahrerspezifische Analyse des Fahrverhaltens zur Parametrierung aktiver Sicherheitssysteme; 4te Tagung Sicherheit durch Fahrerassistenz, München, April 15–16, 2010.

3-20   Gonter, M., Bauer, C., Rojas, R. Fahrerspezifische Prädiktion von Fahrmanövern mit FuzzyLogic; 3. Berliner Fachtagung Fahrermodellierung, Berlin, June 17–18, 2010.

3-21   Schroven, F. Probabilistische Situationsanalyse für eine adaptive automatisiert Fahrzeuglängsführung, Ph. D. Thesis, TU Braunschweig, 2011.

3-22 Wiacek, C., Najm, W. Driver/Vehicle Characteristics in Rear-End Precrash Scenarios Based on the General Estimates System (GES), SAE Technical Paper 1999-01-0817, 1999.

3-23 Schmidt, C. Untersuchungen zu letztmöglichen Ausweichmanövern für stehende und bewegte Hindernisse. 3. FAS-Workshop, 2005.

3-24 Bauer, C. A driver specific maneuver prediction model based on fuzzy logic, FU Berlin, 2012.

3-25 König, M., Gonter, M. et. al. A Sensor System for Pre-Crash Deployment with Extremely Low False Alarm Rate, 5th International Workshop on Intelligent Transportation (WIT), Hamburg, Germany, 2008.

3-26 Waldt, N., Gonter, M., et. al. Testing of Pre-Crash-Airbag-Systems with Extreme Low False Alarm Rate; 9th International Airbag Symposium on Car Occupant Safety Systems, 2008.

3-27 Enzweiler, Kanter, Gavrila. Monocular Pedestrian Recognition Using Motion Parallax, 2008 IEEE Intelligent Vehicles Symposium, Eindhoven, The Netherlands, June 4–6, 2008.

3-28 Nedevschi, S., Danescu, R., Pocol, C., Meinecke, M. Stereo Image Processing for ADAS and Pre-Crash Systems, International Workshop on Intelligent Transportation WIT, Hamburg, Germany, March 18–19, 2008.

3-29 Meinecke, M., König, M., Gonter, M. Multi-Feature Walking Pedestrian Detection Using Dense Stero and Motion, 4th Internaterianl Workshop on Intelligent Transportation (WIT), Hamburg, Germany, 2007.

3-30 Rollmann, G. Frequency Requirements for Automotive Radar—Impact of SARA, International Workshop on Intelligent Transportation WIT, Hamburg, Germany, March 18–19, 2008.

Chapter 4
Functions of Integrated Safety

4.1 Precrash safety

4.1.1 Definition of the precrash phase

Precrash safety systems have to be deployed before the crash occurs. This requires a sensing system that is able to detect all critical objects around the vehicle and is able to measure their distance and dynamics. Depending on the actuator to be used, it is critical to know the type of object (e.g., vehicle, pedestrian, bicyclist, etc.) that is triggering the crash. There are actuators that need information about the collision partner type, because they have to be triggered only in specific situations, or they have to be triggered in a specific manner, or the trigger has to be prevented. Precrash applications are designed to trigger safety measures milliseconds before the collision occurs. Different strategies are currently under consideration including reversible and non-reversible deployment. The reversible activation can take place based on a concrete actuator or warning that occurs early in the precrash phase (e.g., seat-belt pretensioner, optical or acoustical warning). The non-reversible actuators are triggered when the situation becomes so critical that the collision is unavoidable. In this case, a non-reversible actuator like a precrash occupant airbag could be used to improve safety. For technical reasons current precrash safety systems do not always have relevant information available to them. Consequently, intervention only takes place once a collision has been reliably predicted and cannot be avoided no matter what actions the driver may take. This time period before the collision is described as the precrash phase (see Figure 4.1 [4–1]).

Accident evaluation shows that accidents become unavoidable, in terms of driving dynamics, very shortly before the collision. This means that there are only a few hundreds of milliseconds available in the pre-crash phase before the collision occurs (Figure 4.2). During this time range braking could be used to reduce the collision speed or restraint systems triggered to reduce collision impact. Warning devices could also interact with the vehicle occupants during this early stage [4-2]. With additional information from environment and driver modeling, future

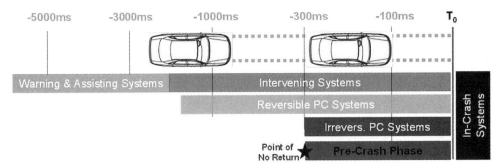

Figure 4.1: Definition of the precrash phase [4-1].

pre-crash systems might be able to intervene sooner, and consequently, further improve safety.

To maximize the availability of a function, environment perception must be very precise in object representation. The function demands initial detection at the earliest possible point, stable target tracking, and short prediction time [4-2]. With regard to the actuators used in a pre-crash function, the activation time should be shorter than the predicted time remaining until the crash to optimally protect the vehicle occupants.

4.1.2 Automatic brake intervention

The GIDAS accident database shows that just more than half of all car drivers apply the brakes in frontal accidents. In this case, the brake assistant provides support. In the remaining accidents where the driver does not respond, only a few hundred milliseconds remain for intervention until the moment of collision. A precrash time period of 100 to 300 milliseconds (ms) allows the application of new safety systems

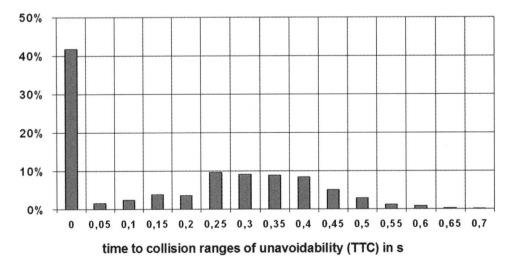

Figure 4.2: Time ranges of unavoidability in terms of driving dynamics in frontal accidents involving cars [4-3].

and also optimizes the layout of existing safety measures. Some of these accidents can be addressed using an automatic emergency brake based on ESC. Due to the response time of the ESC unit, the time required to build up the brake pressure to full braking pressure is too slow at up to 0.6 seconds, which means that rapid brake actuators are required in the precrash phase. Future high-speed electric or pyrotechnic brake actuators can achieve braking gradients up to 1000 bar per second (Figure 4.3).

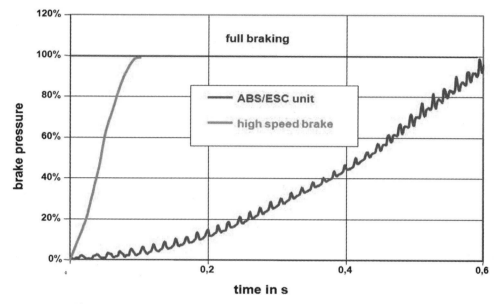

Figure 4.3: Time ranges of accident unavoidability in terms of driving dynamics with cars [4-3].

This means that speed reductions of 1.8 to 4.9 mph (3 to 8 km/h) can be achieved within 100 to 300 ms before the crash to reduce the accident severity. The rapid increase in brake pressure can be particularly advantageous in driving situations where rapid target detection and tracking by the sensors is required, such as when merging onto a road or turning at intersections, as well as when pedestrians or cyclists are crossing the road (Figure 4.4). Precrash sensors recognize that a collision is unavoidable, although these accidents can be detected only very late in the precrash time period.

For this reason, the precrash sensor system must allow optimized object recognition including fast and flexible target identification and target tracking. The activation of rapid brake actuators requires a high level of triggering certainty and an extremely low false alarm rate. In order for a high-speed brake to be triggered, it is necessary to guarantee that the system functions reliably. It is a great challenge to achieve these simultaneous objectives: very high triggering quality and very low false alarm probability.

One example of precrash pyrotechnic brake actuators is the PyroBrake, an economical, small, and lightweight pyrotechnically driven piston unit mounted on the ABS

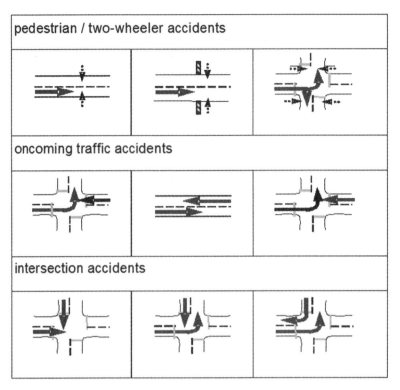

| pedestrian / two-wheeler accidents |
| oncoming traffic accidents |
| intersection accidents |

Figure 4.4: Examples of typical situations in which high-speed brake actuator systems are used.

device (Figure 4.5). The PyroBrake can achieve braking gradients up to 1000 bar per second. The system can initiate emergency braking within 80 ms and can reduce the impact velocity by 3.1 mph (5 km/h) on average [4-4].

Figure 4.5: ABS with PyroBrake unit [4-4].

4.1.3   Irreversible restraint systems

In the case of precrash triggering of restraint systems, there are new developments in airbag configuration compared to the current standard airbags. Triggering the airbag before a collision makes it possible to adapt the airbag to the time profile of the occupant kinematics. When the airbag is triggered before the collision, it will fully inflate when the vehicle occupant begins to move forward in response to the collision (Figure 4.6). The longer inflation time available for the airbag can also be used to increase the volume of the precrash airbag compared to the standard airbag. This early interaction of the occupant with the airbag and participation in the vehicle deceleration, transfers the restraint effect from the seat belt to the airbag, providing support over a larger body area and improving the restraint effect of the occupant in the seat. It may be possible for an occupant who is leaning forward to be pushed back gently. The larger airbag that has begun to deploy during the precrash phase may also improve safety performance in case of an offset impact.

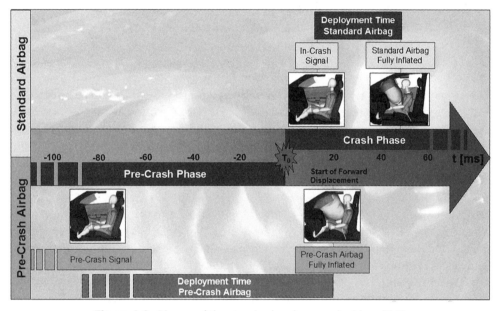

Figure 4.6: Phases of the standard and precrash airbag [4-1].

The aggressive force of the airbag is reduced because of its longer inflation time; this feature is also necessary to satisfy the out-of-position requirements with the larger precrash airbag. To improve interaction of the occupant with the airbag, the volume of the standard airbag can be increased by about 30%, as shown in Figure 4.7 [4-1].

From today's perspective, precrash sensors do not adequately detect collisions, consequently the concept of an irreversible precrash restraint system must provide in-crash triggering as a fall-back level. The precrash airbag consists of a bi-volume bag, the volume and unfolding properties of which can be controlled by means of two arrester straps in the bag.

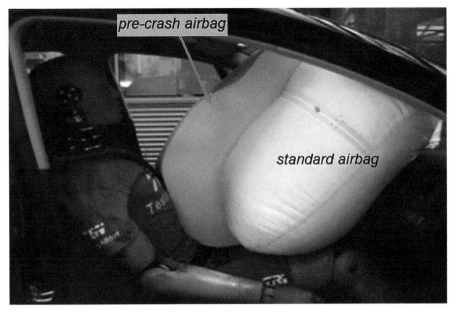

Figure 4.7: Volumes of the standard and precrash airbags [4-1].

The target population for a precrash restraint system that would protect the driver and the front-seat passenger has been researched [4-1]. Accounting for the fact that not all accidents can be detected with sufficient reliability, there was a potential benefit for one third of all car occupants.

4.1.4 Side precrash system

Currently there are different approaches to provide high occupant safety in case of a side impact crash. Today's cars are equipped with highly sophisticated safety measures, which are primarily passive measures, like rigid mechanical constructions of the passenger doors. Experts recognize that the structural enhancements in vehicles in recent years have increased the safety of vehicle occupants. Nevertheless, permanent research activities strive to save as many lives and decrease the injury level as much as possible. In this context a new approach was developed. During a predicted unavoidable collision the vehicle body could be lifted [4-5] enhancing safety for the vehicle occupants because of the shifted point of impact. A radar sensor observes the side area of a vehicle, detects incoming objects, and measures their distance and radial velocity. If an object's trajectory is directed towards the vehicle and becomes dangerous, the radar system deploys applicable safety measure as illustrated in Figure 4.8. The goal is to enhance the occupant's safety.

In the U.S. one third of all vehicles are SUVs, an extremely high market penetration. In Europe the number of SUVs on the road is lower than in the U.S., but is steadily increasing. Collisions between standard passenger cars and SUVs are sometimes very critical because of the different heights of the bumpers. If an SUV hits a standard car at its doors, the occupants may be badly injured, because the bumper of the SUV is relatively high compared to the standard height of a car bumper. The

Figure 4.8: Dangerous situation for a SUV collision vehicle versus a passenger car [4-5].

differences between the standard height of a passenger car's bumper and bumpers of a light duty truck are depicted in Figure 4.9.

Figure 4.9: Bumper heights of a light duty truck versus a passenger car.

When a hazard from a side impact between a car and an SUV is detected, the height of the car is increased by 3.9 inches (100 mm) within 300 ms to avoid the impact between the SUV's bumper and the car door, dramatically decreasing the mechanical load to the occupants. For detecting obstacles and other critical objects a 24 GHz radar sensor is used. The antenna diagram was designed to cover the area of interest for typical side-crash scenarios. Figure 4.10 provides a diagram of the antenna [4-5].

The sensor is able to detect objects of interest like passenger cars, vans, and trucks. This is the basis for a trajectory estimation and risk assessment. When a trajectory is directed to the side area of the vehicle, the chassis is activated automatically lifting the body of the vehicle by 3.9 inches (100 mm) within 300 ms (Figure 4.11) [4-5].

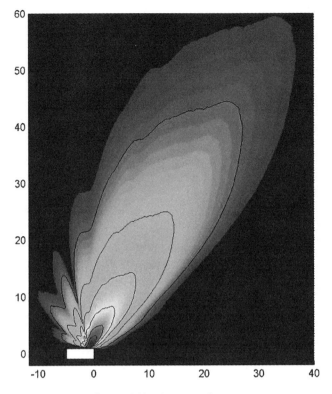

Figure 4.10: Antenna diagram.

From a mechanical point of view, lifting the body is very challenging because of the extremely short action time, approximately 300 ms; the high mass of the body; and high lifting distance of 3.9 inches (100 mm). Special modifications of the chassis are necessary. If the air springs (Figure 4.12) are connected to air pressure vessels via valves, the valves open and the air in the vessel is applied to the air spring when the deployment signal is received. Then the air springs move out of the spring struts raising the vehicle. A similar concept can be implemented using hydraulics.

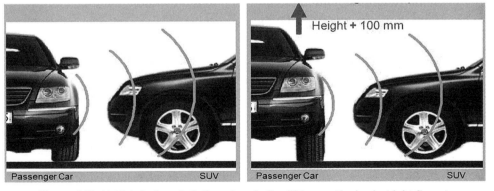

Figure 4.11: Vehicle before (left figure) and after lifting up the body (right figure).

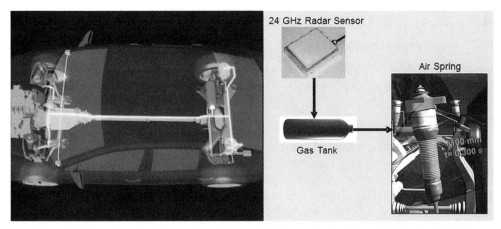

Figure 4.12: Scheme of the activated chassis (left) and activation of Air Spring (right).

This chassis raising mechanism shifts the impact point of the SUV bumper from the car door to the sill board, the stiff portion below the door, protecting the occupants of the vehicle. A crash test performed under IIHS test conditions shows this positive effect of this technology (Figure 4.13) [4-5]. The intrusion in the area of the B pillar is a very important value for vehicle safety. Using the automatic height adaptation device (orange bars) to reduce intrusions is compared to the intrusion of a standard vehicle (blue bars). At the three measurement points, "upper door," "door middle," and "H-point" the test reduced intrusions between 19% and 24%, relevant values for the vehicle occupants. The mitigation of intrusion is compensated by an increased intrusion at the sill beam, which has no direct negative effect on the driver.

The injury level of occupants is decreased as are the loads applied to the human body measured by using SID II dummies in the driver's seat and in back seat behind the driver. The total loads were reduced by approximately 29%.

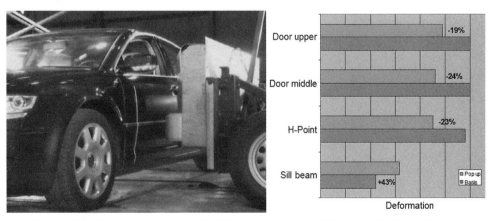

Figure 4.13: Vehicle crash test with IIHS barrier speed of 50 kph (left). Right: Reduced intrusions in the area of B pillar (blue: standard; orange: automatic height adaptation device).

4.2 Systems to integrate avoidance and mitigation

4.2.1 Preventative occupant protection

Reversible restraint systems are triggered before a collision to prepare the vehicle's occupants for a possible collision. The preventative occupant protection system [Figure 4.14] further increases the already high passive protection potential of the vehicle.

If the system detects a potential accident, then the safety belts are tightened electrically. The driver and front seat passenger are fixed in their seats. This allows the airbag and belt system to provide the best possible protection. In addition, if powerful lateral dynamic movements are also detected, the side windows and sliding roof are closed, With the windows closed, head/side airbags can be provided with optimum support, again providing the best possible protection.

The significant feature of this occupant protection system is that it links active and passive safety elements together. The sensors in the driving dynamics control systems such as the brake assistant and ESC are the technical basis for this linkage. They detect critical potential accident situations at an early stage. The system is activated in case of emergency or during very fast brake pedal activation, which usually goes hand-in-hand with activation of the brake assistant, or in case of unstable driving situations such as high understeering or oversteering, with ESC intervention.

In accordance with an integrated safety approach, the preventative occupant protection system is connected to the surround sensors of the ACC. Using the radar

Figure 4.14: Preventative occupant protections [4-6].

sensors and camera technology that can monitor the area around the vehicle, it is possible to warn the driver of an impending rear-end collision. It is also possible to intervene in the braking system and assist the driver with autonomous emergency braking. Using the reversible belt tensioner in conjunction with the emergency braking system effectively reduces the initial forward movement of the vehicle occupants. The front-seat passenger can experience initial forward movement during braking deceleration, whereas the driver can brace himself or herself against the steering wheel.

4.2.2 Integral pedestrian protection

The motivation for integral pedestrian protection is driven by the high proportion of pedestrians who are killed or seriously injured on the roads. In Germany in 2006 pedestrians accounted for 14% of fatalities. In more than 80% of the cases, the fatal accidents are caused by misjudgement on the part of the pedestrians and drivers who reacted too late to avert a collision.

Passive pedestrian protection measures can reduce the injuries suffered by pedestrians in case of a collision. The measures are concentrated in the front area of the vehicle (i.e., pedestrian protection foam and deformable hoods). Passive measures are limited, however. Head injuries from the secondary impact of the pedestrian and the surface of the road can still occur. Consequently, the future objective is to significantly reduce the collision speed by active braking of the vehicle, which promises to reduce pedestrian injuries (Figure 4.15).

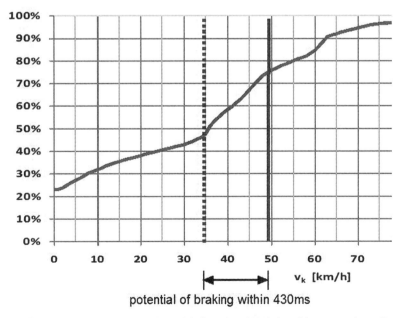

potential of braking within 430ms

Figure 4.15: Pedestrian injury risk function MAIS 2+ with cars and small commercial vehicles as counterparts in the collision.

Accidents between vehicles and pedestrians can be avoided until shortly before the collision if the pedestrian simply stops. This means it is necessary to have systems that intervene quickly to prevent collisions. A typical road situation is shown in Figure 4.16.

Figure 4.16: Typical road situation related to pedestrian safety [4-7].

Predicting a pedestrian's movement is complex. The pedestrian's predicted trajectory is the basis for a deployment algorithm that triggers the alert signal or safety measure in/on the vehicle [4-7]. The higher speed, significant mass, and low contact surface means that vehicles have inertia; and consequently, their continued movement can be described well in physical terms. However, pedestrian movement is much more dynamic. Consequently, a special pedestrian algorithm with a variable safety range is required to represent the movement of the pedestrian (Figure 4.17) [4-8].

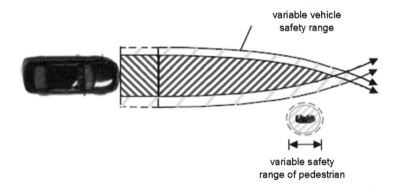

Figure 4.17: Pedestrian algorithms with variable safety range [4-8].

Because of pedestrian dynamics it is not always possible to avoid an accident by using automatic intervention systems such as emergency brakes. Consequently, it is necessary to have an integral pedestrian protection concept that provides information and warnings to the driver in critical situations and only intervenes with automatic braking as a last resort (Figure 4.18).

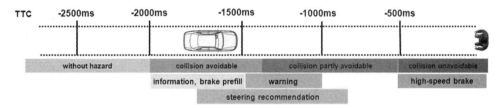

Figure 4.18: Time sequence of the integrated action concept of pedestrian protection.

In the future integrating the various systems into one integrated function with different escalation levels will be necessary throughout the entire accident sequence. A decider will transfer the different individual functions into an overall function. In this way, the driver can first be provided with vectored information about the pedestrians who are in his or her driving corridor or moving towards it, indicating that a collision might be possible. In parallel, the brakes will be pre-filled. Because of pedestrian movement, it is only possible to provide information at a time to collision of less than 2 seconds to avoid sending mis-information to the driver.

Depending on the predicted impact position of the pedestrian with the vehicle, the decider located at the top of the decision-making process activates a haptic steering recommendation (or superimposed steering) to the steering wheel if a slight steering reaction may be sufficient to avoid the collision. In other cases, a brake warning consisting of a brake jolt and an optical/audible warning is created, to which the driver should respond intuitively. This could be followed by automatic emergency brake intervention if the driver has not responded and the possible collision has not been avoided up to this point (approximately less than 500 ms). In this precrash phase, it would be advantageous to have a high-speed brake to keep the residual accident severity as low as possible (Figure 4.19).

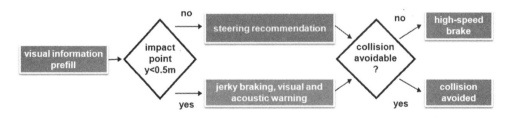

Figure 4.19: Example of an integrated action concept [4-9].

4.2.3 From steering support to automated driving intervention

As the previous chapter on integrated pedestrian protection has described, pedestrian protection involves a different integrated safety approach. With the functionality of steer-recommendation and steer-assistance it is also possible to avoid critical situations with pedestrians (see also [4-10]). The modular decider described is also usable in this case. Independent of the velocity and distance of the pedestrians, the decider can calculate the assist strategy. One requirement is the unavoidability of the collision. Figure 4.20 shows the context between the last point of steering and braking. For low velocity collisions braking is the best option to lower the distance and time to collision. At higher velocities steering is necessary to avoid a collision because of the higher braking distance independent of the square of velocity.

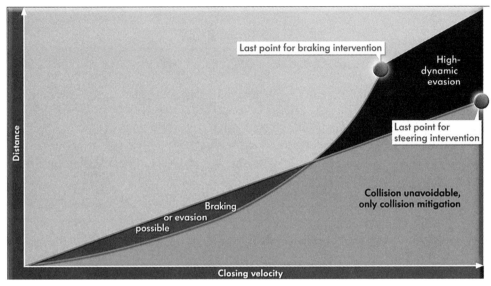

Figure 4.20: Braking and steering possibilities to avoid a collision as a function of velocity and distance.

In the future the functionality of steering support will be further developed into an automated evasion change maneuver. It will be one of the challenges of automated driving, because automated driving means accident free driving. Automated driving can be divided into three segments: partly automated, highly automated, and fully automated driving. During fully automated driving the driver does not need to supervise or take control. Highly automated driving means that the function works for a limited duration and the driver will have to resume control in due time. The first step in bringing highly automated driving to the public could be the emergency stop function with automated cooperative lane change where the driver's sudden medical emergencies can cause severe accidents.

In the future the occupants could use the time gained during automated driving for other things like reading the news via an internet connection, working or resting. In this case accident free driving means that the integrated safety functions will support the driving functions to avoid all early stage risks. In this case cooperative driving is necessary with soft steering and braking interventions, otherwise it will be a challenge to achieve high occupant acceptance of a highly automated driving mode.

4.2.4 Rescue and recovery

In addition to systems that avoid accidents and minimize their consequences, the integrated safety approach also includes support for vehicle occupants after the crash. These systems are grouped together under the heading of rescue and recovery. They include keeping the lighting systems operating and providing structures that support rescue or indicate cutting marks which make it easier for rescue services to position their rescue equipment in an optimum way.

Integral systems will be able to improve rescue scenarios in the future by using data from the pre-accident and accident phases. In this way, important information such as the number of occupants, severity of the accident, and the accident type will be provided to the rescue personnel before they arrive on the scene to allow the rescue to be organized appropriately.

Traffic accidents occur at all times of the day and night, as well as in cities, in the country, or on the highways. Often accidents involve a single driver. Because of this, there may be no eye-witness to quickly call for assistance for the injured. Often, the injured person cannot call for help either, because he or she is trapped inside the vehicle, unconscious or his or her mobile phone has been destroyed in the accident.

Rapid or timely medical assistance is a key factor in survival. For example, the chance of surviving a cardiac arrest caused by an accident drops by 10% per minute [4-11]. The automated system ECall could provide assistance here by automatically informing the rescue services in case of an accident or providing a guaranteed communication pathway for the injured person. ECall is linked to the airbag control unit where it can be triggered by the activated airbag, but it can also be activated manually by the driver (Figure 4.21). The system uses a crash proof box equipped with a GPS receiver and global system for mobile communication (GSM) module.

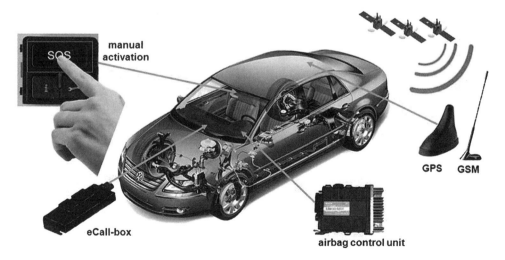

Figure 4.21: ECall components in principle.

If it is activated, a minimum set of data is transferred to the rescue services. These data include location, chassis number, and timestamp thereby making it easier for the rescue personnel to reach the accident and allowing a rapid estimate of the situation. Implementing ECall is based on many European standards and directives, including ETSI (European Telecommunication Standards Institute) and CEN (European Committee for Standardization). Specification of a uniform ECall system, as well as its corresponding infrastructure and server structure, has not yet been completed on a European level.

4.2.5 Development process of integral functions

Because of the interdisciplinary nature of the work, development of integrated safety functions requires significant collaboration during the development process. Areas of responsibility for the chassis, electronics, body, and safety systems are merging to an ever-increasing extent (Figure 4.22).

Parallel development of networked functions and systems requires a particular structure in the development tasks and processes. Starting from the function specification, systems and components must be developed in a shared architectural process. In this case, the roles of those involved in the development process must be distributed in a defined manner, and it is also necessary to ensure that the documentation of the specifications and interfaces needed for the function is provided clearly and transparently. Last but not least, success of the development process requires that tests and trials be performed precisely.

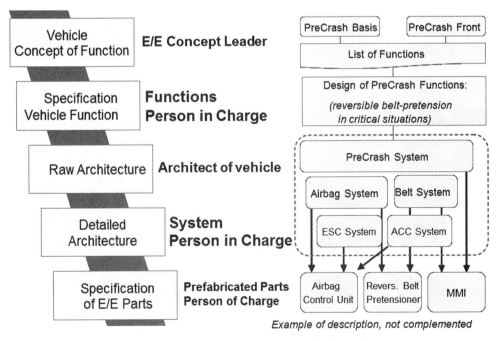

Figure 4.22: Networked developing process of electronic systems (example of precrash occupant protection).

4.3 Car-to-infrastructure safety

The goal in the development of new safety systems is to reduce the number of people killed or injured in traffic accidents. Using new technologies like vehicle communication systems can support this goal. Car-to-infrastructure safety describes the development of safety functions in this area and forms the basis for future accident-free driving when linked with integrated safety.

4.3.1 Introduction

Using car-to-car communication or car-to-infrastructure communication, it is possible to further improve vehicle safety. The European Union announced its objective to halve the number of road casualties within the next decade. The automotive industry supports this challenge as part of its vision zero, the aim of which is to avoid any accident victims. Active safety systems are one part of this vision as are innovative new sensor technologies.

Vehicle communication systems could be useful in different ways (Figure 4.23). The undeniable advantage of sensors is the ability to see around corners and through obstacles and to get advance warning of potential dangers. Systems for better traffic flow and efficient driving can be designed to react to traffic light circuits or on pre-knowledge of the traffic situation.

Sensors could warn drivers of accident hotspots like traffic jams or immobilized vehicles. Existing safety systems are not designed to detect and warn drivers about such situations [4-12]. Intersection assistance systems could also benefit from data exchange between cars which are not visible to each other, reducing intersection and turning accidents. Safety systems based on environmental sensors are limited to detect crossing vehicles at intersections. This is partially due to the limited range of the sensors and obstruction of the field of view by trees and buildings. When vehicles communicate with each other it is possible to obtain information on their position and velocity as well as on their weight, structure stiffness, or state of their airbags. Thus, adaptive safety functions could be developed based on vehicle communication and fusion of environmental sensors. The potential for enhanced

Figure 4.23: Vehicle communication systems with a foresight.

safety of the different vehicle-to-vehicle systems in Germany, based on the GIDAS database, is shown in Figure 4.24 [4-13].

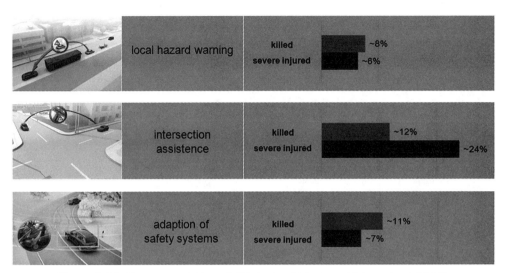

Figure 4.24: Target population of different vehicle-to-vehicle systems [4-13].

4.3.2 Car-to-car based functions and requirements

The potential for information exchange between cars offers new ways to develop safety functions. Communication based functions (car-to-car functions) require a minimum number of equipped cars to provide a useful degree of effectiveness. To reach this number and to ensure a successful market introduction, German car manufacturers, as well as suppliers and other partners, are working together in research projects like SIM-TD (SichereIntelligenteMobilität-Testfeld Deutschland) and standardization groups like the car-to-car communication consortium. The main activities of these groups are the standardization of the communication process and the evaluation of possible functions under real life conditions.

An additional challenge for the introduction of such communication based functions is to define the most efficient applications to be supplied for a target population with low market penetration. Therefore, some exemplary safety functions and criteria for their selection are explained in the following section. An important proposition for the description and evaluation of car-to-car functions for safety issues is the correlation between the functional range and the number of cars which are equipped with the needed communication hardware. Furthermore, an important source for application ideas are accident analysis and statistics, which can be used to find equivalent functions to prevent the most common hazard situations. These application ideas have to be combined with the available technical solutions.

The V2V-Communication is a completely new sensor system. Using a highly dynamic VANET (Vehicular ad-hoc Network), vehicles share data with other vehicles in the traffic stream. Through this sensor, ignoring several restrictions, there is the opportunity to transmit amounts of information concerning the current vehicle state of driving and potentially dangerous road conditions, if several restrictions are neglected.

Analyses by the U.S. Department of Transportation's NHTSA acknowledge that connected vehicle technology could potentially impact 76% of vehicle crash types involving non-impaired drivers in the U.S. [4-9]. Specifically, NHTSA research shows that these technologies could help prevent a majority of crashes that typically occur in the real world such as crashes at intersections or while vehicles are switching lanes. Research projects like Safety Pilot from Intelligent Transportation Systems Joint Program Office emphasize the importance of safety issues for the design and evaluation of vehicle-to-vehicle systems. These research projects will probably influence the National Highway Safety Traffic Administration's regulation processes. The technologies being tested include in-car collision warnings, alerts that a vehicle ahead has stopped suddenly, intersection collision warning, emergency electronic brake light, and other similar safety messages. The information collected from the safety pilot test will be used by NHTSA to determine whether to proceed with additional vehicle–to-vehicle communication activities by 2013 [4-14].

The first generation of car-to-car communication for safety issues will focus on warnings like the visible warning symbols in a head unit as seen in Figure 4.25 and information about hazard areas. This means, for example, that it is possible to develop functions that warn about the dangers of black ice, oil spills, traffic jams, accidents, broken down vehicles, and emergency service vehicles (Figure 4.26).

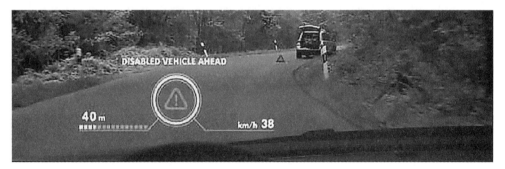

Figure 4.25: Example of a visible warning of a car-to-car hazard situation [4-15].

Depending on the vehicle speed and time needed to track and warn drivers about a possible hazard, a certain communication time range should be identified to allow for timely information exchange. A vehicle traveling 124 mph (200km/h) is driving on a highway toward the last cars in a traffic jam. The first point of information exchange should be at a communication range of 820.2 feet (250m) so that 4.5 seconds are available for tracking and warning. A second point for discussion is the position accuracy needed during the warning function. For a traffic jam warning function detection of individual lanes should be not necessary; detection of the highway in general should be sufficient. The longitudinal accuracy of the position should be around ± 16.4 feet (5m). Therefore both the informing time as well as the warning time can be influenced, but at an acceptable level.

Figure 4.26: Examples for a car-to-car-function which aims to give other drivers warning of a hazard situation [4-13] [4-15].

For other warning functions like intersection movement assist, do not pass warning, or forward collision warning the sensor might need to detect individual lanes. Then the lateral accuracy of the relative position should be around 1.6–3.2 feet (0.5–1 m) [4-12]. The Safety Pilot project is investigating different strategies for vehicle related path prediction and path history to achieve this ambitious requirement [4-16].

The transfer time of the data is another point to mention. For safety functions the transfer time is one of the most important criteria. It defines the duration between the occurrence of a hazard situation and the transfer of information to relevant traffic participants. When the transfer time is shorter, there are more opportunities for the system to react. The transfer time includes the time to establish a connection, transfer the data, and close the connection if necessary. Referring to the described low-level safety functions, a transfer time of less than one second is needed and sufficient at the same time. The data exchange is another point to consider. For all functions based on communication information the exchange has to be secure to protect the privacy of the drivers and to prevent possible system manipulation.

4.3.3 Automatic braking intervention by vehicle-to-vehicle and sensor fusion

As described, the application of active safety systems is a worthwhile approach to reduce the number of traffic accidents. With vehicle-to-vehicle communication or vehicle-to-infrastructure communication it is possible to optimize the potential of vehicle safety. Therefore, diverse systems like hazard-warning functions could be developed. For time-critical functions like an automatic braking system, it is necessary to fuse car-to-infrastructure information and environmental sensors.

New functions, like avoidance of intersection accidents, could be developed by using sensor fusion. System reaction and speed have to be based on exact position data of the surrounding vehicles. The accuracy of the position should be less than ± 3.2 feet (1m). For safety applications the currently used GPS data is not accurate enough. An approach to fuse vehicle-to-vehicle data with an additional environmental sensor (radar, camera, etc.) is proposed to resolve the conflict between high functional requirements and resulting hardware costs. Using this approach it is feasible to realize vehicle-to-vehicle safety applications with considerably lower hardware requirements. Junction assistance systems could benefit from data exchange, particularly if the surround sensors in the crossing vehicle cannot detect the other vehicle because it is outside the line of sight. In the case of time-critical functions with automatic intervention, it is necessary to link car-to-infrastructure information with data from the surround sensors. The advantages of this fusion involve additional validation of the surround sensor data and early-stage system intervention because the track time of the surround sensors is reduced. The fusion creates an expanded environmental model. A sample scenario of a fusion of car-to-infrastructure information with data from surround sensors is shown in Figure 4.27. Without using vehicle communication, the surround sensor only detects the crossing vehicle when it establishes direct visual contact. This is followed by the track time. With vehicle communication the first data exchange takes place much earlier. It is possible for the surround sensors and the safety system to be pre-conditioned by reducing tracking time. In this case, data can be passed to the safety system at an earlier point, including additional data such as the vehicle mass for passive safety systems.

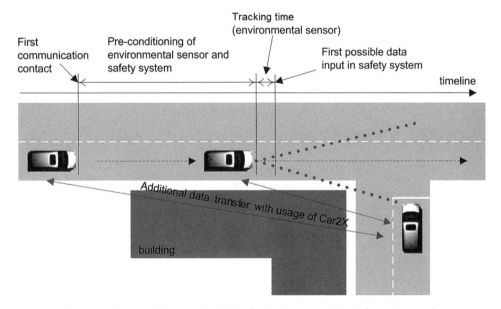

Figure 4.27: Sample scenario of the fusion between C2X information and data from the surround sensors [4-17].

The same situation is given in a curve scenario (Figure 4.28). The fusion of the environmental sensors and car-to-car communication data reduces the tracking time. The early automated braking intervention may allow the vehicle to avoid the accident.

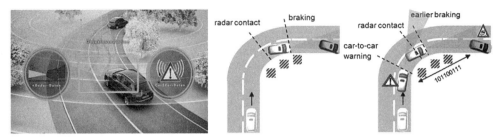

Figure 4.28: Fusion of the environmental sensors and car-to-car communication data leads to a reduction of tracking time and early automated braking intervention.

To fuse the additional communication data it is useful to look at the normal sensor input of an active safety system. To build on the established approach of fusing different environmental sensors to get a surrounding model of the environment of the vehicle, the car-to-infrastructure data is added to this model. With communication data and the existing models of the surroundings the CarMatching algorithm creates an extended model. These new data could be used as inputs for certain safety applications. The extended architecture is shown in Figure 4.29.

The different requirements of the safety functions, as well as the effectiveness of such functions concerning the accident statistics and the potential of crashworthiness (target population from GIDAS database), leads to a conceivable time line, which might be used for introducing such functions to the market [4-7]. With an increasing rate of market penetration for the communication units the efficiency of the systems is expected to increase (Figure 4.30). However, security issues concerning car-to-infrastructure safety will need to be addressed.

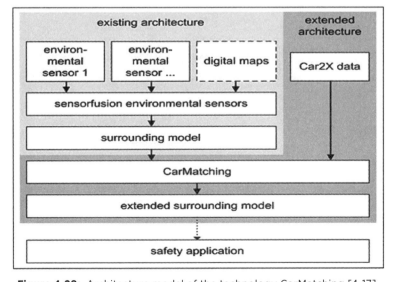

Figure 4.29: Architecture model of the technology CarMatching [4-17].

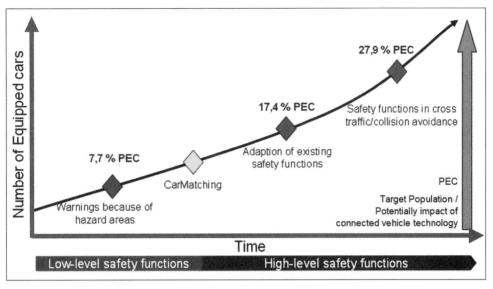

Figure 4.30: Target population of low and high-level vehicle-to-vehicle safety functions [4-17].

4.3.4 Cooperative driving

In the future, when the market saturation of cars equipped with car-to-x technology is higher and the fusion with other sensors is in place, another generation of safety functions with vehicle-to-vehicle communication can move forward. These functions can to target additional hazard situations. The existing safety and assistance functions, which are based on environmental sensors, can be adapted to have a greater functional range and improve safety and comfort using driving assistance systems based on supplemental car-to-car information. The vision is not only to mitigate or avoid a collision, but to avoid risks that arise in an early stage because of cooperative driving and with car-to-car communication. Figure 4.31 provides one example of cooperative driving behavior in a lane changing situation, the acceleration lane on a highway. The information base (e.g., set turn signal or

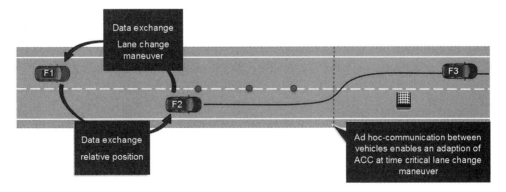

Figure 4.31: Example of cooperative driving behavior in lane changing situations.

lane change maneuver data), can be used to allow for an earlier ACC brake activation in reaction to a lead vehicle that is changing lanes. This can result in higher consumer acceptance.

One car could obtain surrounding sensor information from the other car and create a warning if the car's driver attempts to change lanes in unsafe conditions. The sharing of surrounding sensor information by vehicle-to-vehicle communication could contribute to wider vehicle safety in the future.

4.4 References

4-1 Wohllebe, T., Gonter, M. et. al. Potential of Pre-Crash Restraints in Frontal Collisions; 8th International Airbag Symposium on Car Occupant Safety Systems, December 2006.

4-2 Müller, M. Meinecke, M. Gonter, et. al. Signal Processing Strategies for a Multi Sensor Pre-Crash Application; 22. Internationale VDI/VW-Gemeinschaftstagung Integrierte Sicherheit und Fahrerassistenz, Wolfsburg, Germany, October 2006.

4-3 Gonter, M. "PyroBrake baut noch 100 Millisekunden vor dem Crash Energie ab," Pressetag Volkswagen, Wolfsburg, Germany, June 2008.

4-4 Waldt, N., Gonter, M., et. al. Testing of Pre-Crash-Airbag-Systems with Extreme Low False Alarm Rate; 9th International Airbag Symposium on Car Occupant Safety Systems, 2008.

4-5 Meinecke, M., Gonter, M., Wohllebe, T., et. al. Side-Pre-Crash Sensing System for Automatic Vehicle Height Level Adaptation; 3rd International Workshop on Intelligent Transportation (WIT), Hamburg, Germany, 2006.

4-6 http://www.volkswagen.de/vwcms/master_public/virtualmaster/de3/meta content/Technik_Lexikon/proaktives_insassenschutzsystem.popup.html (October 2010).

4-7 Meinecke, M., Gonter, M., Widmann, U., Current Trends in Vehicle Active Safety and Driver Assistance Development; FISITA, World Automotive Congress, Munich, Germany, September 14–19, 2008.

4-8 Seiffert, U. et. al. Handbuch Kraftfahrzeugtechnik; Fahrzeugsicherheit Kap. 9, Seite 763-804, 6. Auflage, ViewegVerlag, September 2011.

4-9 Research and Innovative Technology Administration, Connected Vehicle Applications. Online, June, 2011, U.S. Department of Transportation.

4-10 Eckert, A., Hartmann, B., Sevenich, M., Rieth, P., Emergency steer and brake assist—a systematic approach for systems; ESV Conference, Paper Number 11-0111.

4-11 http://www.avd.de/startseite/service-news/news/alle-news/2009/oktober/avd/avd-and-kvda-fordern-warnsystem-fuer-einsatzfahrten-der-rettungskraefte.

4-12 Haak, U., Sasse, A., Hecker, P. On the Definition of Lane Accuracy for Vehicle Positioning Systems. Intelligent Autonomous Vehicles, Volume 7 Part 1.

4-13 Franke, K., Gonter, M., Küçükay, F., Car2Car Sicherheitsfunktionen—High Safety–Low Cost; VDI-Tagung Fahrzeugsicherheit Fokus Elektromobilität, Berlin, Germany, October 5–6, 2011.

4-14 U.S. Department of Transportation, News NHTSA 18-11, www.dot.gov/briefing-room, Orlando, Oct 19, 2011.

4-15 www.car-to-car.org, date 2009-12-08.

4-16 www.its.dot.gov/safety_pilot.

4-17 Rößler, Gonter, M., et. al. Car2X Safety—Future Development of active safety systems based on vehicle communication systems; 7th International Workshop on Intelligent Transportation (WIT), Hamburg, Germany, March 23–24, 2010.

Chapter 5
Biomechanics and Protection Criteria

5.1 Biomechanics

5.1.1 Introduction

More than 50 years ago, researchers in the U.S. and Europe and later in other parts of the world began studying biomechanics and human tolerance during road accidents with greater intensity.

Biomechanics is the synthesis of mechanics and medicine with the purpose of determining the load limits of the human body, its structures, and soft tissue [5-1]. Biomechanics are representative of a macroscopic approach. The material properties of different soft tissues were determined relatively early on at a microscopic level [5-2]. They form the basis for the finite element method (FEM) modeling of the human body.

Important contributions to the analysis of human resistance against "impact loading" were made by the American Colonel Stapp, who was the first person to subject himself to deceleration from a speed of 632 miles per hour (mph) (around 1000 km/h) to zero mph in 1.4 seconds. For more information, refer to [5-3].The direct quote was as follows:

"The Stapp Car Crash Conferences are named in honor of Colonel John Stapp. USAF (MC), who pioneered (and is still pioneering) in establishing human impact tolerance levels. His historic rocket sled rides at Holloman Air Force Base, New Mexico, in 1954, in which he voluntarily subjected himself to up to 40g accelerations while stopping from a speed of 632 miles per hour in 1.4 seconds, still represent the best basis for quantifying human tolerance to acceleration. In addition to his own dangerous volunteer work, he has directed countless other safety research programs involving human volunteers, animals and cadavers. The equipment and techniques developed under his guidance have become standard in this research area and have contributed much to the advancement of safety. The naming of these conferences after Colonel Stapp is a fitting tribute to a man who has dedicated his life—even

to the point of risking it—to research aimed at increasing man's chances of survival in adverse crash environments.

The conferences were initiated at the University of Minnesota (Colonel Stapp's alma mater) under the able direction of Professor James J. Ryan, another outstanding researcher in crash safety. For four years, the conferences were held at either the University of Minnesota or an appropriate U.S. Air Force base. Currently, the conference rotates annually among four sponsors: The University of Minnesota (1961), the United States Air Force (1962), the University of California at Los Angeles (1963), and Wayne State University (1964). The 1965 meeting will again be held at the University of Minnesota on October 20, 21, and 22. The proceedings of the conference are published in bound from and will, it is hoped, become a valuable reference source."

Subsequently, research results were shared and discussed during the annual Stapp Conferences, the International Research Committee on the Biomechanics of Impacts (IRCOBI) [5-4] and the European Experimental Vehicle Committee (EEVC), all very important institutions that create and exchange biomechanics research results. For automotive safety the biomechanical data are a vital tool to quantify the human body's critical tolerance levels.

Understanding of biomechanics data has made it possible to define a number of requirements for the development of vehicles and their components. These requirements form the basis for the protection criteria, which are recorded as physical parameters within the corresponding test equipment. The applicable legal requirements define the measured or calculated data that should not be exceeded.

5.1.2 Tolerance limits

Human tolerance limits describe fractures, organ damage, and other injuries. A classification system is provided by the Abbreviated Injury Scale (AIS) and Overall Abbreviated injury Scale (OAIS). AIS and OAIS judge both single and total injuries and have a scale from 0 to 6, as shown in Table 5.1 [5-5].

The tolerance limits are influenced by age, sex, anthropometric data, mass, and mass distribution. A typical injury description is given in Table 5.2.

Table 5.1: AIS Abbreviated Injury Score	
0	Not injured
1	Minor
2	Moderate
3	Serious
4	Severe
5	Critical
6	Maximum
9	NFS (Not Further Specified)

			Abdomen and		Extremities and
AIS	Head	Thorax	Pelvic Contents	Spine	Bony Pelvis
1	Headache or dizziness	Single rib fracture	Abdominal wall; superficial laceration	Acute strain (no fracture or dislocation)	Toe fracture
2	Unconscious <1 hr; linear fracture	2–3 rib fracture; sternum facture	Spleen, kidney, or liver; laceration or contusion	Minor fracture without any cord involvement	Tibia, pelvis, or patella; simple fracture
3	Unconscious 1–6 hours; depressed fracture	≥4 rib fracture; 2–3 rib fracture with hemothorax or pneumothorax	Spleen or kidney; major laceration	Ruptured disc with nerve root damage	Knee dislocation; femur fracture
4	Unconscious 6–24 hours; open fracture	≥4 rib fracture with hemothorax or pneumothorax; fail chest	Liver; major laceration	Incomplete cord syndrome	Amputation or crush above knee; pelvis crush (closed)
5	Unconscious >24 hours; large hematoma (100 cc)	Aorta laceration (partial transection)	Kidney, liver, or colon rupture	Quadriplegia	Pelvis crush (open)

Table 5.2: AIS Examples by Body Region [5-6]

Because of the inherent risk of injury, it is nearly impossible to use human beings in accident simulation tests. Therefore, with the help of simulation tools such as anthropomorphic dummies or mathematical models, a wide range of subjects (vehicle occupants, pedestrians) can be considered. Cadaver testing was very important in the past. Potential injury limits volunteer testing.

5.1.3 External injuries

In the past, lacerations of the face and neck could occur caused by impact with the vehicle's windshield. A rating of these injuries was developed by Professor Larry Patrick, Wayne State University in Detroit, Michigan. The bonded and laminated windshield did contribute significantly to the reduction of lacerations and helped to keep occupants and airbags in defined parts of the vehicle interior.

Skull fractures caused by head impact against the vehicle interior are described by Swearingen [5-7] as follows:

The deceleration of the head multiplied by the mass of the head results in a force, which creates a fracture of the part of the head with a load per surface unit.

Acceleration limits are 200 g for the face, 30 g for the nose, and 40 g for the chin. Chest injuries could occur through impact with the steering column and the dashboard. The impact force should be lower than 5000 Newtons (N). The chest deformation should not exceed 1.9 inches (5 cm).

A fracture of the upper leg could occur if longitudinal forces are higher than 11,000 [N], as shown in Figure 5.1.

Since the head and torso have been well protected from injury during collisions by the combination of three point safety belts and air bags, injuries to lower extremities are gaining more attention.

Body Part	Mechanical Variables	Load Values
Total Body	$a_{x\,max}$ $\bar{a}_x$	40...80g 40...45g, 160...220 ms
Brain	$a_{x\,max}$, $a_{y\,max}$	100...300g WSU-curve with 60g, T>45 ms 1800...7500 rad/s²
Skull Fracture	$a_{x\,max}$, $a_{y\,max}$	80...300g depending on the size of the impact area
Forehead	$a_{x\,max}$ F_x	120...200g 4000-6000 N
Cervical Spine	$a_{x\,max\,thorax}$ $a_{y\,max\,thorax}$ F_x $\alpha_{max\,forward}$ $\alpha_{max\,rearward}$	30...40g 15...18g 1200...2600 N shear force 80°...100° 80°...90°
Thorax	$a_{x\,max}$ F_x S_x	40...60g, t>3 ms 60g, t<3 ms 4000...8000 N 5...6 cm
Pelvis-Femur	F_x $A_{y\,amx}$	6400...12500 N force application in the femur 50...80g (pelvic)
Tibia	F_x E_x M_x	2500...5000 N 150...210 Nm 120...170 Nm

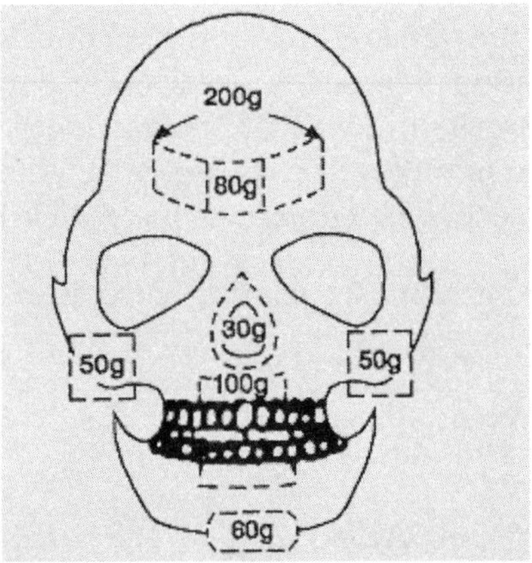

Figure 5.1: Biomechanical limits on humans [5-7].

5.1.4 Internal injuries

It is much more difficult to assess internal injuries. The greatest challenges are the loading of the brain and the cervical spine.

The head should not exceed an acceleration of 80 g in an anterior-posterior direction for more than 3 ms. Concussion and severe brain damages are judged by the Patrick diagram [5-8], shown in Figure 5.2. The data show the correlation between the deceleration level and exposure time.

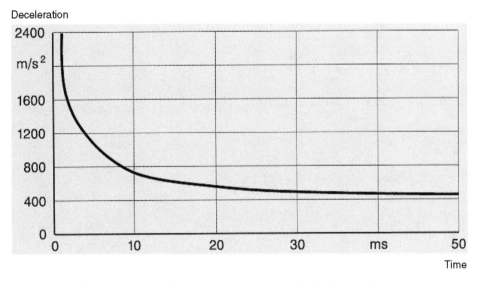

Figure 5.2: Patrick tolerance curve related to the human brain.

The Patrick curve was also used as a basis for the head-injury-criteria (HIC).

The tolerance for rotational acceleration was investigated by Fiala [5-9]. According to his data a rotational acceleration of more than 7500 rad/s^2 (radian per second squared) at a brain mass of 1300 g is critical in causing head injuries.

Like head injuries neck injuries are also important. As the connecting element between body and head, the neck and spine may be damaged in all type of accidents. This is especially true for the neck, which is highly stressed by the relative forward movement of the head to the body (flexion) and the relative forward movement to the body (extension). Depending on the muscles involved and the specific characteristics of the occupant, severe injuries can occur. Excessive movement around the occipital condyle is extremely critical.

Injuries of the cervical spine became more important because safety belts and airbags already provide significant occupant protection. The safety belt force limiter helps to reduce chest injuries, although for different occupant weights, sizes, and ages an adaptive solution offers great promise.

5.2   Protection criteria

Because humans cannot participate in tests where injuries could occur, a large number of test devices are available in the form of both hardware and as mathematical models. In contrast to human biomechanical tolerance data, specific tolerance data related to these test devices are required for development projects.

The frontal impact tolerance limits, shown in Figure 5.3, are universally valid, whereas country specific requirements show some differences.

The HIC, HPC ~ head injury/protection criteria are defined as follows:

$$HIC \leq 1000 \approx \left[\frac{1}{t_2 - t_2)} \int_{t_1}^{t_2} a_{res}\, dt \right]^{2,5} (t_2 - t_1)$$

The relevant time periods are 15 or 36ms.

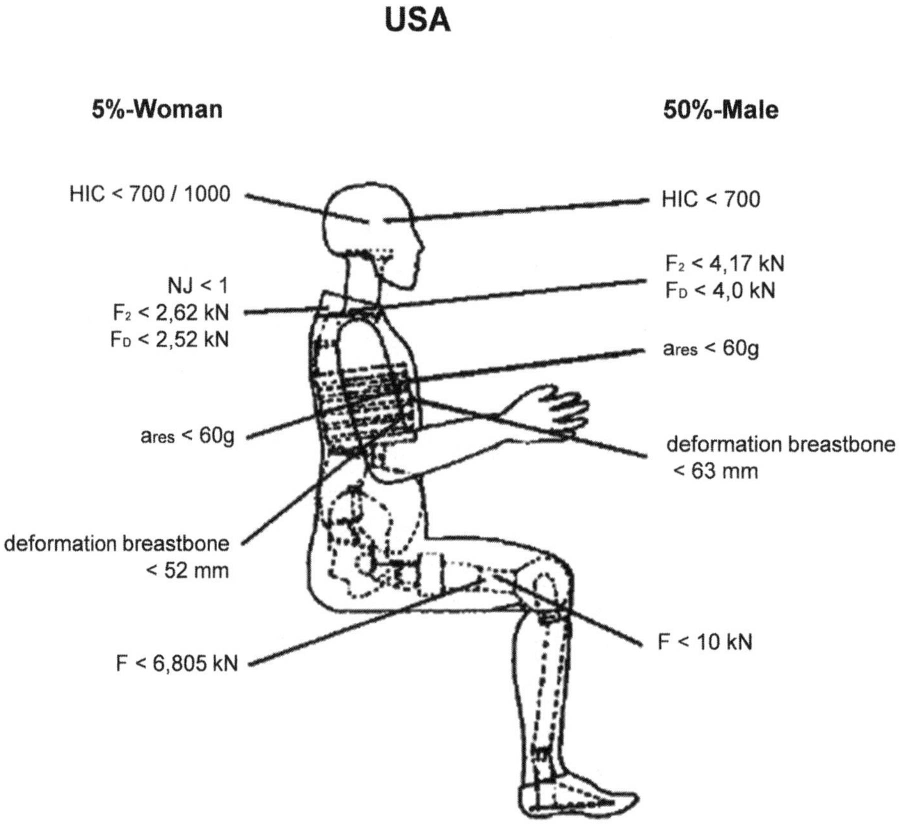

USA

5%-Woman   **50%-Male**

HIC < 700 / 1000   HIC < 700

NJ < 1
F₂ < 2,62 kN   F₂ < 4,17 kN
F_D < 2,52 kN   F_D < 4,0 kN

                                                a_res < 60g

a_res < 60g

                                                deformation breastbone
                                                       < 63 mm

deformation breastbone
       < 52 mm

                                                F < 10 kN
F < 6,805 kN

Figure 5.3: Legal requirements for tests with dummies.

Most tolerance limits can be taken from the Figure 5.3 although some should be explained more in detail.

- Upper femur force < 9,07 KN at 0ms, < 7,56KN after 10ms
- Tibia index (TI), measured at a upper and lower point should be lower than 1,3[-]

The tibia index is calculated as follows:

$$TI \quad = \quad \left| M_R / (M_c)_R \right| + \left| F_z / (F_c)_z \right|$$

M_x =   Bending moment about the x-axis

M_y =   Bending moment about the y-axis

$(M_c)_R$ =   Critical bending moment and shall be lower than 225 Nm

F_z =   Compressive axial force in the z direction

$(F_c)_z$ =   Critical compressive force in the z direction and shall be lower than 35,9 KN

$$M_R \quad = \quad \sqrt{(M_x)^2 + (M_y)^2}$$

The tibia index is calculated for the top and bottom of each tibia; however, F_z may be measured at either location. The value obtained is used for the top and bottom TI calculations. Moments M_x and M_y are measured separately at both locations.

For neck injuries, the neck injury criteria (NIC) should not exceed tensile and shear forces KN as function of time f(t).The torque around the y-axis should in the rearward motion below 57 Nm and in the forward motion below 190 Nm.

For frontal crashes, the steering wheel displacement should be below 3.1 inches (80 mm) in the vertical plane and below 3.9 inches (100mm) in the horizontal plane. No precise criteria have been defined for injuries to the feet; however, in general the foot should not be subjected to excessively high torque, and severe impacts with the vehicle interior should be avoided.

For lateral impact the tolerance levels shown in Figure 5.4 should be considered.

The abbreviations in the figure have the following meaning:

RDC = Rip deflection criteria

APF = Abdomen loading

PSP = Symphysis loading

In general, overall vehicle design and configuration should take total system tolerances, vehicle restraint systems, and measurement devices into consideration. This means that the development goals must be much more stringent for the production models. Dummies and simulation measurement devices are installed for the head, neck, chest, pelvis, and lower extremities. Acceleration forces, deflection, and torque are usually measured.

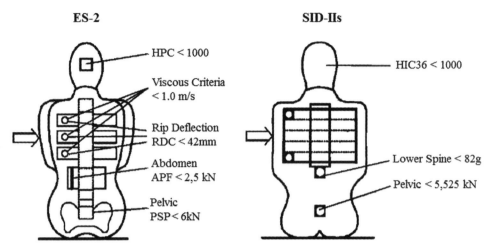

Figure 5.4: Requirements for dummies related to lateral impact.

		Frontal Impact Dummies			Side Impact Dummies				Rear Impact Dummies		Child Dummies				
		HIII 50%	HIII 5%	HIII 95%	ES-2	ES-2re	SID-IIs	World SID	HIII 50%	BioRID II	Crabi	Cami	HIII	P Series	Q Series
Europe	ECE-R94	•													
	ECE-R95				•										
	ECE-R44													•	3)
	Euro NCAP	•	5)		•			5)		•				•	1)
USA	FMVSS208		•								•		•		
	FMVSS214					•	•	2)							
	FMVSS213										•	•	•	•	
	FMVSS202a								•						
	US-NCAP	•	•			•	•								
	IIHS	•					•			•					
Asia	Japan Trias 47 (1-4)	•			•										
	JNCAP	•	•		•					•					
	China Gesetz	•			•										
	China NCAP	•	•		•		•			•				•	
	Korean NCAP	•			•					•					
AUS	ADR (Frontal, Side)	•			•										
	Australian NCAP	•			•					4)				•	

1) planned for 2013 (Q18m, Q3) respectively 2015 (Q6, Q10)
2) planned for 2016
3) planned for 2014
4) from 2012
5) planned for 2015

SafetyWissen by carhs.

Figure 5.5: Dummies used in different test modes and countries [5-10].

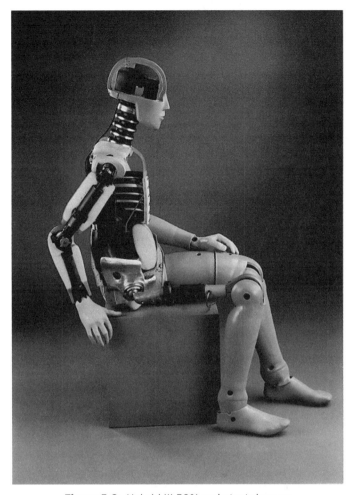

Figure 5.6: Hybrid III 50% male test dummy.

The variation in dummies used is relatively high and worldwide uniformity in dummy models remains a dream. Figure 5.5 [5-8] shows an overview of dummies used in different countries and for different tests.

In the following figures some examples of the dummies used for the dynamic testing are shown. Figure 5.6 depicts a 50 percent male dummy for frontal impacts, Figure 5.7 shows two dummies used in side impacts, and Figure 5.8 illustrates the Bio-RID, a tool for rear impact evaluations.

Not only are three dimensional hardware dummies available in various sizes, but a number of test devices for the head, the legs, and other parts of the body are also available. Several mathematical models are also available for use in the development process.

Figure 5.7: Side dummy ES2 [5-11].

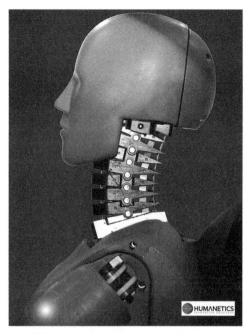

Figure 5.8: Bio-RID dummy used in rear impacts [5-11, 5-12].

5.3   References

5-1   Nahum, A.M. et. al. Accidental injury (Biomechanics and Prevention) Springer, New York 1993, ISBN 3-540-97881-x.

5-2   Yamada, H. Strength of Biological Materials, Williams & Wilkins 1970.

5-3   Proceedings of the 8th Stapp Car Crash Conference, Detroit 1964.

5-4   IRCOBI proceedings 1998, International Conference on Biomechanics of Impact, Göteburg, Sept. 1998.

5-5   The Abbreviated Injury Scale—2005 update 2006, Association for the Advancement of Automotive Medicine, Des Plaines, IL.

5-6   Pike, J. *Automotive Safety*, Society of Automotive Engineers, Warrendale, PA, 1990.

5-7   Swaeringen, J.J. Tolerance of the Human Face to Crash Impact, Federal Aviation Agency, July 1965, Report No. AM65-20, Oklahoma City, OK.

5-8   Patrick, L. "Human Tolerance to Impact—Basis for Safety Design," SAE Paper No. 650171, Society of Automotive Engineers, Warrendale, PA.

5-9   Fiala, E., *et al.* "Verletzungsmechanik der Halswirbelsäule, Forschungsbericht der Technischen Universität, Berlin 1970.

5-10   Carhs, www.carhs.de.

5-11   HUMANETICS www.humanetics.eu.

5-12   Kelly, James R. BioRID-Ilc Rear Impact Crash Test Dummy, Denton Inc., USA.

Chapter 6
Mitigation of Injuries

In the following chapter the areas in the field of mitigation of injuries related to vehicle design are described.

6.1 Quasi-static test requirements on the body in white

Body in white is the unpainted vehicle's basic frame/unit structure, including sheet metal but minus bolt-on-components

6.1.1 Tests on seats and seat belt anchorage points

When the inner seat belt latch is connected at the seat, the seat and the safety belt anchorage point are tested simultaneously. Using rigid body blocks in accordance with the FMVSS 210 [6-1], the pulling force has to be equally applied. For each body block a resistance of more than 14.000 N must be accomplished. The local reinforcements in the upper part of the B-pillar must be designed in such a way as to avoid tearing the pillar because the local rigidity is too high.

The seats in the standard design cannot withstand the required forces for seat belt anchorage points because the seat itself must only resist forces 20 times the seat curb weight for 30ms. Therefore, the inner anchorage point is connected to the stable middle tunnel of the body in white via a serrated seat rail. Seats where the upper anchorage point is mounted to the seat back must withstand the torque and force applied by the seat belt. For this reason they are relatively heavy, because of the rigidity that is required.

6.1.2 Roof strength

For the evaluation of the roof strength the FMVSS 216 and 216 a [6-2] require a quasi-static test for vehicles with a total mass lower than 10,132 pounds (4596 kg). The vehicles are tested via a steel plate with a crush velocity of ≤ 0.51 inches (13mm) per second. The applied force has to be three times the vehicle's empty weight for vehicles with a mass of less than 6001 pounds (2722 kg), and 1.5 times the vehicle's empty weight for vehicles with a mass equal to or greater than 6001 pounds (2722 kg). The deformation of the roof has to be less than 5 inches (127 mm), and the contact force equal to or less than 222 N with a 50% male in the seating position. The phase in period for vehicles with an empty weight of less than 6001 pounds (2722 kg) started on September 1, 2012, and ends on September 1, 2015.

6.1.3 Side structures

Besides the dynamic tests there is also a quasi-static test that can be conducted in accordance with the FMSS 214 [6-3]. In this test, a cylinder is pushed vertically into the longitudinal axis along the side of the test vehicle. The lower part of the cylinder has to be 5 inches (12.5 cm) above the lowest point of the door. The test device is 12 inches (30.5 cm) in diameter. It is long enough that it extends 0.5 inches (1.27 cm) higher than the bottom edge of the side window. Figure 6.1 shows the resistance of a body in white with a door with and without side beams. In these tests the maximum resistance force is limited by the capability to transfer the load via the hinges and door latches to the A, B, and C-pillars.

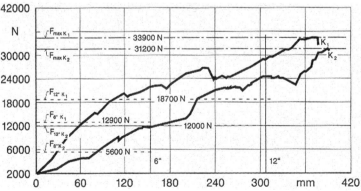

Figure 6.1: Test results of quasi static door impact tests (with and without door beams).

As the result of this, the maximum resistance force reached is approximately the same for a door with and without a side beam. The differences can be seen in the beginning of the deformation at 6 and 12 inches (15.2 and 30.5 cm), where the side beams show their positive safety performance.

6.2   Dynamic simulation of vehicle collisions

6.2.1   Frontal collision

To evaluate possible damage to head lamps, side markers, and other important components, relatively low velocity impact tests are performed against a fixed wall and with a pendulum against the vehicle bumpers. Where the U.S. has legislation in place to regulate these tests, in Europe this is handled primarily by the relevant insurance classes. In such cases the repair costs are more important. At impact speeds up to 5 mph (8 km/h) no damages should occur. To determine the repair cost the speed of the vehicle is increased to 9.9 mph (16 km/h). Since 2010, a bumper test shown in Figure 6.2 was gradually introduced in addition to the existing tests.

Figure 6.2: AZT-Allianz Centre for technical tests.

At higher speed the kinetic energy of the collision must be absorbed to a high degree by the deformation of the vehicle components.

The deceleration, velocity, and deformation time in a frontal collision against a fixed barrier (which is located 90° to the vehicle's longitudinal axis) is shown in Figure 6.3.

In the rebound, shown by the negative velocity, the elastic deformation is around 10%. This results in a velocity change of approximately 34 mph (55 km/h) at an initial impact speed of 31 mph (50 km/h).

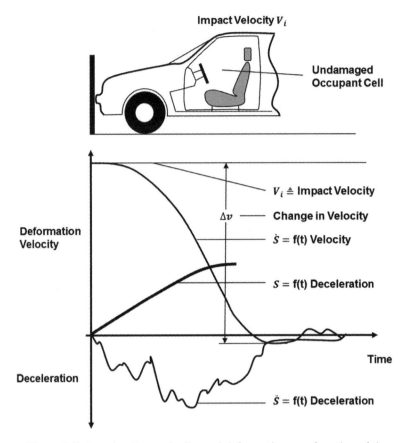

Figure 6.3: Deceleration, velocity, and deformation as a function of time.

It is interesting to analyze the percentage of the energy absorption in a frontal crash for the various body in white components. Figure 6.4 gives an overview for a standard body in white.

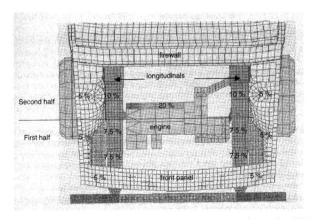

Figure 6.4: Distribution for energy absorption during a frontal collision [6-4].

The impact against a fixed wall can be described under the assumption of a plastic impact using the following parameters:

$\ddot{S}_v$ = deceleration of the vehicle during the impact as f(t)

$\dot{S}_v$ = velocity of the vehicle during the impact as f(t)

S_v = deformation of the vehicle during the impact as f(t)

V_i = impact velocity, Δv = change of velocity

$F, \bar{F}$ = Deformation force, average deformation force

Under the assumption that the deformation force is constant, the data are as follows:

$$\ddot{S}_V = -a,\; \dot{S}_{V(t=0)} = v_i,\; S_{V(t=0)} = 0$$

$$\ddot{S}_{V(t)} = -a \text{ and } \frac{m_v}{2}\left(v_i^2 - \dot{s}^2\right) = \int_0^s F \cdot \Delta s = \bar{F}s_V$$

$$\dot{S}_{V(t)} = -a \cdot t + v_i$$

$$S_{V(t)} = -\frac{a \cdot t^2}{2} + v_i \cdot t$$

(6.1)

$$S_{V(s)} = \frac{v_i^2 - \dot{S}^2}{2a}$$

(6.2)

If one replaces the average deformation force by:

$$\bar{F} = m_V \cdot a \text{ and } v_i^2 = 2a \cdot S_V$$

We get the following equation:

$$\frac{m_V}{2} v_i^2 = m_V \cdot a \cdot s_v \text{ or } v_i^2 = 2a \cdot s_V$$

From the formulas above we can conclude, that in a frontal collision, the deformation forces for different vehicles are different as well. With a higher vehicle mass we also get a higher deformation force:

$$F = m_v \cdot a.$$

(6.3)

It is also evident, that with an equal deceleration level, which is important for the occupant load, the deformation length during a collision against a fixed barrier is velocity and not mass dependent. At an impact speed of around 31 mph (50 km/h) the majority of the vehicles have a deformation length between 17.7 to 29.5 inches (450 to 750 mm). At impact speeds of approximately 35 mph (56 km/h) this deformation length increases by around 3.9 to 5.9 inches (100 to 150 mm).

The behavior during an impact against a 30° barrier correlates relatively strongly with a vehicle to vehicle crash. If we compare this with a frontal impact (90°) against a fixed barrier, the occupant loading is lower. In contradiction the offset impact creates a much higher loading of the vehicle structure, because the major part of the absorbed energy has to be taken by one side of the vehicle. The respective velocity changes in the center of gravity Δv_x and Δv_y relative to the center of the vehicle are:

30° – impact $v_i \sim 50$ km/h, $\Delta v_x = 57{,}6$ km/h, $\Delta v_y = 14{,}4$ km/h

40% – offset impact $v_i = 50$ km/h, $\Delta v_x = 52{,}5$ km/h, $\Delta v_y = 13{,}7$ km/h

In a frontal collision of two vehicles with the masses m_1 and m_2 the following relationships are valid:

$$\frac{m_1 \cdot v_{i1}^2}{2} + \frac{m_2 \cdot v_{i2}^2}{2} = \frac{1}{2}(m_1 + m_2) \cdot u^2 + F \cdot (\Delta s_1 + \Delta s_2)$$

with u ~ common velocity after the impact.

$$u = \frac{m_1 \cdot v_{i1} + m_2 \cdot v_{i2}}{m_1 + m_2} \tag{6.4}$$

If we define the relative velocity at the beginning of the impact of both vehicles,

$$v_r = v_{i1} - v_{i2}$$

we get:

$$v_r^2 = 2F(\Delta s_1 + \Delta s_2) \cdot \frac{m_1 + m_2}{m_1 \cdot m_2}$$

The velocity changes of both vehicles are:

$$\Delta v_1 = v_{i1} - u = (v_{i1} - v_{i2})\frac{m_2}{m_1 + m_2} = v_r \frac{m_2}{m_1 + m_2} \tag{6.5}$$

$$\Delta v_2 = v_{i2} - u = (v_{i2} - v_{i1})\frac{m_1}{m_1 + m_2} = v_r \frac{m_1}{m_1 + m_2} \tag{6.6}$$

The question is often asked: which velocities have to be taken to simulate a frontal collision of two identical vehicles. The equation shows that the velocity changes in both types of collisions have to be identical ($v_{i1} = v_{i2}$). Therefore, the change of velocity against the fixed barrier Δv_{1B} has to be equal for the vehicle to vehicle collision Δv_{1Fz}. For the plastic impact the following equations are:

Case I: Impact against a fixed barrier

$$v_{1B} = (v_{i1} - v_{i2}) \frac{m_2}{m_1 + m_2}$$

with $m_2 = \infty$ and $v_{i2} = barrier = 0$

$$\Delta v_{1B} = v_{i1}$$

Case II: Impact against a second vehicle

$$v_{1Fz} = (v_{i1} - v_{i2}) \frac{m_2}{m_1 + m_2}$$

with $m_2 = m_1$ and $v_{i2} = -v_1$

$$\Delta v_{1Fz} = v_{i1}$$

Therefore the impact of a vehicle with v_{i1} against a fixed barrier matches a vehicle to vehicle collision where each vehicle has a speed of v_{i1}, although in opposite directions. Or it is equal to an impact with two times v_{i1} into another identical vehicle that is not moving. If we assume a plastic impact at 31 mph (50 km/h), impact against a fixed barrier has, therefore, the same consequences of a collision between two identical vehicles with an equal mass and a relative velocity of 62 mph (100 km/h). If we take into consideration the fact that the elastic deformation is higher in a fixed barrier collision, then the relative velocity in a vehicle to vehicle crash should be around 68.3 mph (110 km/h). Due to the rebound phase in the fixed barrier crash, the equivalent test speed against the fixed barrier is around 10% lower than real world accidents.

The safety simulation tests for frontal collisions are very numerous. Besides the test defined by the OEMs, the number of tests increased significantly over the last years due to the New Car Assessment Program (NCAP) and legal requirements. One example is the different U.S. test requirements defined by the FMVSS 208, shown in Figure 6.5 [6-5].

6.2.2 Lateral collisions

Like the frontal collision tests, a number of tests based on legal requirements and consumer information tests (NCAP) are used to improve vehicle safety in lateral collisions.

In the FMVSS 214 [6-3] test a moving barrier drives into the side of the tested vehicle in a crabbed configuration. Two types of dummies are used, the ES-2re in the front and the SID-IIs in the rear at the impact side. Figure 6.6 shows the layout of the FMVSS 214 test.

In Europe, a moving barrier weighing 2094 pounds (950 kg) (IIHS 1500 kg) drives under 90° at 31 mph (50 km/h) into the side of the tested vehicles during the ECE-R95, 96/27/EG, and the US-IIHS tests. The following dummies are used in these tests: the ES-2 for the ECE/EG test front seat impact side and SIDIIs for the IIHS 2 also on the impact side.

FMVSS 208: Frontal Impact Requirements: In-Position

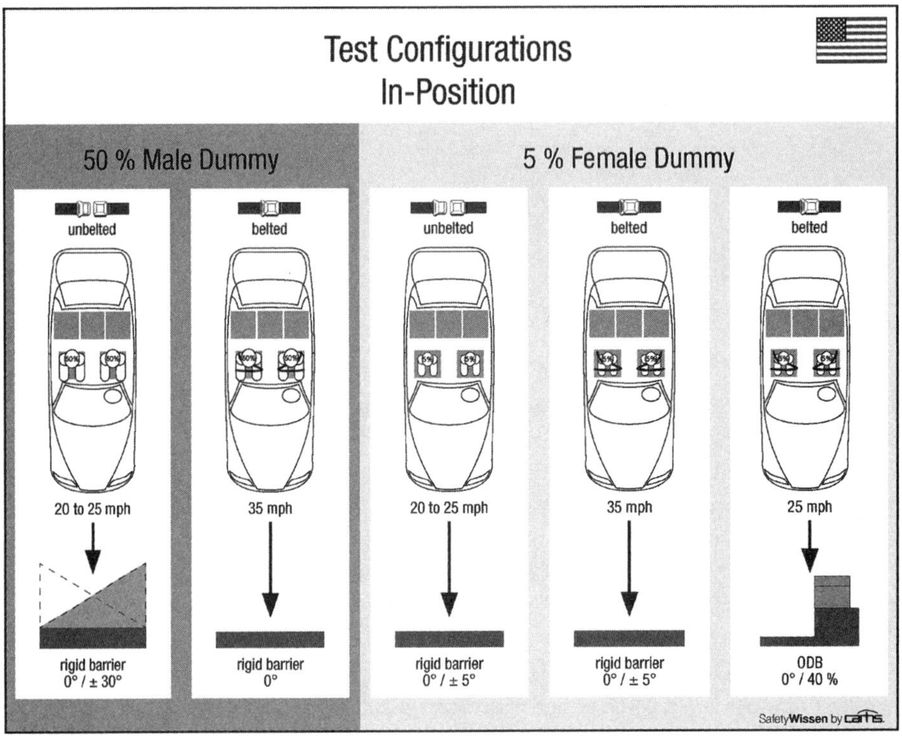

FMVSS 208: Frontal Impact Requirements: Out of Position

Front seat	Dummy	Test configuration
Driver side	HIII 5% female dummy	chin on airbag module in steering wheel
		chin on top of steering wheel
Passenger side	1 year old child dummy	in 23 defined CRS / positions
	3 year old CRABI - dummy	chest on instrument panel
		head on instrument panel
	6 year old CRABI - dummy	chest on instrument panel
		head on instrument panel

SafetyWissen by carhs.

Figure 6.5: Examples for frontal accident simulation tests [6-6].

Pole-side impact tests are defined in the European and U.S. NCAP, the FMVSS 201, and the FMVSS 214. In these tests the vehicle is crashed at speeds between 18 mph (29 km/h) and approximately 20 mph (32 km/h) against a fixed pole with a diameter of 10 inches (254 mm). According to the test procedure the following dummies are used: ES-2 in the EURO NCAP, SID-H3 in the FMVSS 201, SID IIs 5% in the US-NCAP, and SID IIs and ES-2re in the FMVSS 214.

Side Impact Test Procedures

MDB - Side Impact Tests according to FMVSS 214 / US NCAP

Requirement	FMVSS 214 old rule	FMVSS 214 new rule	US NCAP
Impact angle	side 90°, 27° crab angle	side 90°, 27° crab angle	side 90°, 27° crab angle
Impact velocity	54 km/h (48 km/h in 90° direction)	54 km/h (48 km/h in 90° direction)	61 km/h (55 km/h in 90° direction)
Barrier	MDB, 1368 kg Mass, 279 mm above ground, 1676 mm width	MDB, 1368 kg Mass, 279 mm above ground, 1676 mm width	MDB, 1368 kg Mass, 279 mm above ground, 1676 mm width
Dummy	2 DOT-SID	Front seat: ES-2 re / Back seat: SID IIs (Build Level D) (impact side)	Front seat: ES-2 re / Back seat: SID IIs (Build Level D) (impact side)
Protection Criteria	Chest TTI < 85 g (4-doors) Chest TTI < 90 g (2-doors) Pelvis acceleration < 130 g	SID IIs: HIC 36 < 1000 Chest acceleration < 82 g Pelvis force < 5,525 kN ES-2 re: HIC 36 < 1000 Chest deflection < 44 mm Abdominal force < 2,5 kN Pelvis force < 6 kN	
Velocity Vectors for Moving Deformable Barrier	48 km/h / 24 km/h / 54 km/h	48 km/h / 24 km/h / 54 km/h	55 km/h / 27 km/h / 61 km/h

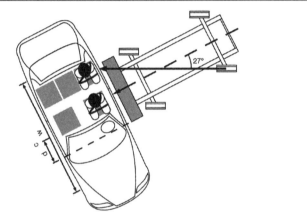

SafetyWissen by carhs.

Figure 6.6: Layout of the FMVSS 214 test [6-6].

6.2.3 Rear-end collisions

Simulating a rear-end collision serves basically two functions:

- To judge the performance of the seats and head rests during rear-end collisions

- To ensure that the fuel tank systems stay intact

As simulation tools different movable barriers (see the description side impact testing) are used. A new requirement is the offset rear-end collision of a 3016 pound (1368 kg) movable barrier traveling approximately 50 mph (80 km/h) into the rear end of the tested vehicle as seen in FMVSS 301 [6-7].

6.2.4 Vehicle rollover

The vehicle rollover accident is getting more attention, especially in the U.S., because of the high number of fatalities from rollovers on American roads. To prevent rollovers installation of ESC is required. To prevent leakage of the fuel should a rollover occur, gravity valves are installed between the fuel tank and the active charcoal filter, which are closed if the vehicle turns around its longitudinal axis. Because of the situation in the United States, several existing standards have been reinforced, for example, the FMVSS 206, 214, and 216.

At the moment the following tests are used to evaluate rollover:

- The evaluation by a dynamic handling test (fish-hook, NHTSA)

- The tightness of the total fuel system using a "RhönWheel" test, which is conducted before and after the frontal, lateral, and rearend collision tests; the vehicle is turned with the "Rhönwheel" in 90° increments, during which the vehicle must remain in one position for at least five minutes

- Test of the complete vehicle with dummies during dynamic roll over tests; for this test the vehicle is put under 23° on a sled, which is decelerated from 30 mph (48 km/h). As a consequence, the vehicle turns as shown in Figure 6.7. It is very interesting, that compared to frontal and lateral collisions the temporal processes are much longer, up to several seconds. They guarantee a much higher chance of survival for the belted occupant without severe injuries, as long as the vehicle compartment stays intact.

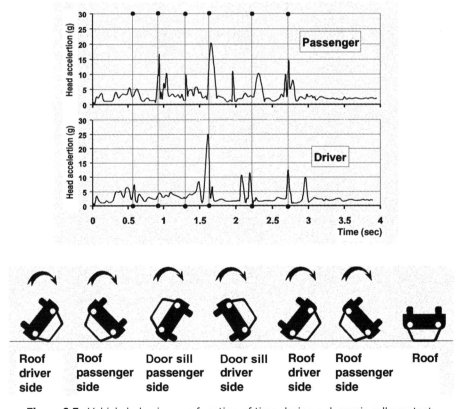

Figure 6.7: Vehicle behavior as a function of time during a dynamic rollover test.

The FMVSS 226 [6-8] is a new test for ejection mitigation. In this test a specifically designed simulated head with a mass of 39.7 pounds (18 kg) is driven against the front and rear side windows at up to four locations at two different speeds, approximately 10 mph (16 km/h) and 12.4 mph (20 km/h). The simulated head should not be ejected by more than 3.9 inches (100 mm) out of the vehicle contour.

6.3   Occupant protection

6.3.1   Vehicle interior

In addition to the classic restraint systems like safety belts and air bags, the total passenger compartment of the vehicle has to meet specific requirements. For example, all interior parts must be designed with a minimum radius of 0.09 to 0.1 inches (2.5 to 3.2 mm) if they are used in a specified head impact area, where head contact is possible. A simulated head with a diameter of 6.5 inches (165 mm) is specified in the requirement. The FMVSS 201 and the ECE R21 require that head deceleration is limited in multiple head impact areas in the vehicle's interior.

The steering column and the airbag steering wheel require special attention. They interact with the seat belt and the seat providing important elements for occupant protection.

Although, the FMVSS 208 has superseded the requirement that the steering wheel not exceed an intrusion in horizontal and vertical directions, the FMVSS 204 is still valid. The defined movement of the steering system's gearbox allows the steering column to telescope. Even with potential future systems like steer by wire, the airbag will be installed in front of the driver.

Like the steering wheel, the foot pedals should not penetrate into the interior of the vehicle and should not shatter into sharp edges during frontal collisions.

All other interior vehicle parts contribute to the overall interior safety as well. The design of the inside door panel is especially important in reducing the occupant loads in lateral crashes.

6.3.2 Restraint systems

Restraint systems can be differentiated between devices that have to be activated for their protective effect like safety belts and child restraints and devices that are automatically activated during the accident like safety belt tensioners, air bags, and belt movement limiters.

Modern seat belt systems demonstrate an excellent protection level for the vehicle occupant in case of an accident, especially when they are combined with air bags. A crucial criterion for the quality and performance of a restraint system is how the individual components match each other. That means that only the perfect interaction of vehicle structure, steering wheel movement, seat performance, interior design, safety belt characteristics, and air bag performance will guarantee optimal occupant protection. Adaptive restraint systems contribute further to the reduction of the occupant load depending on the accident severity and characteristics of the occupants.

6.3.2.1 Safety belts

The three point belt with the automatic retractor is the main protective system used worldwide. The three point belt is normally installed on the outside of the seats. In vehicles with five seats we also find three-point belts for the middle seats. The seat belt latch for movable seats is installed at the seat. The upper outside part of the belt is normally installed at the B and C pillar; in some cases, it is also installed at the upper part of the seat's back. The upper anchorage point has to be in an area specified by legislation (Figure 6.8). Most outside upper anchorage points are height adjustable. The comfortable position for the occupants and its protection function still complies with requirements.

Two independent mechanical systems lock the safety belt during an accident. One mechanism uses the vehicle acceleration or deceleration according to the pendulum principle. At an impulse of more than 0.7 [g] the locking is triggered. The second

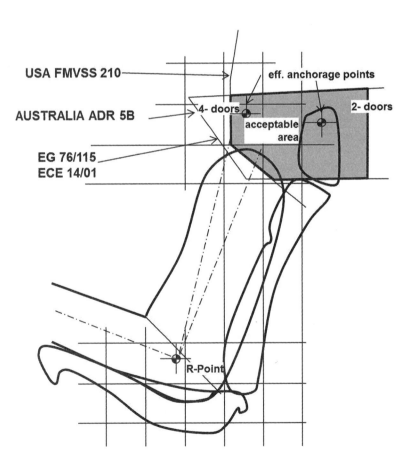

Figure 6.8: Legally defined areas for the upper safety belt anchorage point.

mechanism reacts to the safety belt excerpt acceleration where only a certain belt length can be pulled out of the retractor. This means that above an excerpt acceleration of more than 1 [g], the locking must occur in between 0.78 inches (20 mm). For the optimum protection via the safety belt and vehicle structure, the pyrotechnic belt tensioner became standard equipment. Figure 6.9 shows the layout of a pyrotechnical belt tensioner.

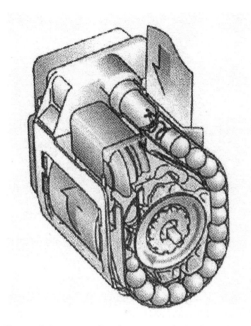

Figure 6.9: Example of a pyrotechnical tensioner [6-9].

After reaching a certain deceleration level, dependent on the accident severity, the pyrotechnical belt tensioner is tightened by sensor controls, in 20 ms; the seat belt is tightened dynamically up to 2 KiloNewtons (KN) and permanently up to 500 N. Pre Safe Systems use electrical motors to tighten the belts.

We also find a high number of force limiters that reduce excessive loads to the occupant's chest.

The type of additional devices used depends on the interaction of all the safety elements involved in the specific accidents, vehicle deformation characteristics, seats, restraint systems, and the geometric layout of the vehicle.

The comfort of the safety belt is also very important because it contributes to the safety belt usage rate. The installation of the safety belt latch at the movable seat and the belt height adjustment, as shown in Figure 6.10 contributed positively to acceptance of wearing safety belts.

Another measure to minimize belt slack is the optimization of the retraction force of the three point automatic belt. The retractor must guarantee perfect retraction if the belt is not in use, and for comfort reasons the retraction force should not be high enough to cause pain at the chest. Sometimes two retractors per seat will be used to increase the comfort in the pelvic belt.

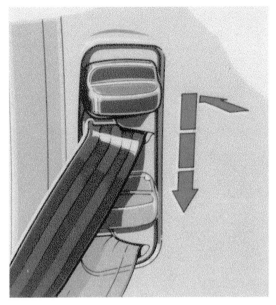

Figure 6.10: Safety belt height adjustments.

6.3.2.2 Child restraints

There is no single child restraint system on the market that covers all ages and stages of child development. The different mass distribution between children and adults must be considered in the design of child restraint systems. Some manufacturers offer child seats which have also an optimal design for installation in the vehicle.

Within the European Union the requirements for child restraints are defined by the ECE-R44 [6-10].

class 0   for children with a body weight of less than 22 lbs (10 kg)

class 0+ for children with a body weight of less than 28.7 lbs (13 kg)

class I   for children with a body weight between 19.8 lbs (9 kg) to 39.7 lbs (18 kg)

class II   for children with a body weight between 33 lbs (15 kg) to 55 lbs (25 kg)

class III for children with a body weight between 48.5 lbs (22 kg) to 79.4 lbs (36 kg)

This classification is also valid for the ISOFIX child restraint systems, which have defined mounting points in the vehicles. The ISOFIX Tether child seats have additional restraint options for the upper mounting. Most vehicle manufacturers take child protection seriously and offer a complete range of child restraint systems.

Figure 6.11 shows the Volkswagen child restraint program for the Golf 6 in Germany [6-11].

Figure 6.11: Child restraint seats for the Golf 6.

There are two different options for seat installation: rearward and forward mounted systems. In the rear facing seat the child's back faces the direction in which the vehicle is moving. This installation is recommended for babies and small children because it provides head support, minimizing the risk of severe neck injuries during frontal collisions. Specifically in this age group the head mass is heavy in relation to the neck muscles that have not fully developed. In vehicles with air bags for the right front seats the air bag is deactivated when the child seat is installed. Older children can use child restraint systems that face forward.

6.3.2.3 Airbags

We will also find different designs for airbag systems which depend on the philosophy of the vehicle manufacturers and the legislation in the U.S. Because of the U.S. requirements for passive restraints, the airbags have to the tested partially with safety belts not in use. This definitely does not increase safety belt usage and could be called a perversion of vehicle occupant safety. The airbag system should be only an amendment to the safety belt. The different occupant dynamics, (e.g., whether the occupant is belted or unbelted) require different airbag systems, each of which need to be adapted to the installed restraint system.

With airbag systems using three point belts, the occupant moves forward in the beginning of a frontal accident relative to the vehicle, until the belt is locked. After that, the relative forward movement continues with a lower relative velocity. This movement is influenced by the pelvic belt which creates a vertical force downward into the seat.

At the end of the crash phase we also find higher head rotation that could be very significantly reduced using the installed driver and passenger airbags. The airbag system is a very compact unit. Several sensors analyze the deceleration-time-function. If the sensor has determined that the accident is severe enough, the igniter in a pyrotechnic box is activated. The gas, created by the pyrotechnic material, fills the front, side, and knee airbags.

The trigger time for the sensor at a 31 mph (50km/h) barrier impact is in a range of 30ms. The time for the bag to inflate is approximately 25ms. The initial interior pressure of the bag is approximately 0.3–0.8 bar. The performance sequence for the driver and passenger side airbags after the sensor is activated is shown in Figure 6.12.

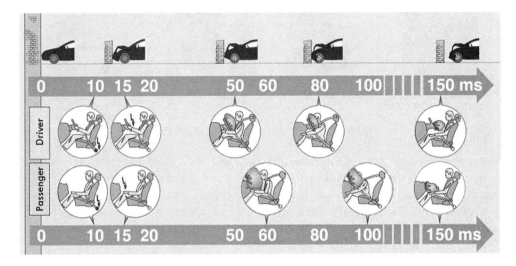

Figure 6.12: Time Sequence and protection function of a restraint system including a three point belt and airbags in a frontal collision (driver and passenger side).

In countries with a low safety belt usage rate, particularly in the U.S., the defined protection criteria for the occupants must also be achieved without factoring in seat belts.

This requires a corresponding design of the area below the dashboard by installing an energy absorbing knee bolster, like those in vehicles with passive belt systems. If the airbag is the only restrain system, the sensor system design has to work much more precisely and must also trigger in oblique impacts (–30° to +30°) relative to the vehicle longitudinal axis. The system has to be designed in such a way that the airbag system provides the total restraint function. The volume of the driver airbag in such cases is around 80 liters (l) and that of the passenger airbag around 150l. The design of the bag has to cover a wide range of occupants in addition to the 50%-male and the 5% female. The system design not only has to cover a smaller occupant; in an extreme case it has to cover children standing in front of the airbags. Very tall and very heavy occupants have to be considered as well. Openings in the bag guarantee a controlled force-deflection-function. For the controlled deployment of the airbag the pressure time has to be varied according to the accident situation, the size of the occupant, and the occupant's seating position.

After the first airbags were installed in production cars there were more than 100 fatal injuries in the U.S., and the regulation had to be totally reworked incorporating requirements for out of position occupants and for occupants of different sizes. Table 6.1 gives an overview of these changes.

Table 6.1: Test requirements for airbag systems in accordance with FMVSS 208					
	Dummy-Type				
Test Requirements	50 % male	5 % female	6 year old child	3 year old child	12 month old baby
Fixed Barrier, belted, 48 km/h, vertical	X	X	N/A	N/A	N/A
Fixed Barrier, belted, 56 km/h, perpendicular	X	N/A	N/A	N/A	N/A
Fixed Barrier, belted 32–40 km/h, vertical and 30° offset	X	N/A (only perpendicular)	N/A	N/A	N/A
Offset, deformable barrier, crash driverside, belted, 40 km/h		X			
Automatic Suppression (Static test to determine when the airbag is automatically deactivated, if a child is using a vehicle seat or if the child is sitting in the passenger seat in positions defined in the standard)	N/A	N/A	X	X	X
Option, inflation with low risk (The vehicle must fulfill the injury criteria as defined in the standard when the driver- and passenger airbag is inflated)	N/A	X	X	X	X

Also the requirements for advanced airbags for vehicles sold in the U.S. were introduced in two steps in 2007 and in 2009. The second step increased the barrier impact speed for tests with belted dummies in a frontal barrier collision test from 29.8 to 34.8 mph (48 to 56 km/h).

In the meantime, airbag systems were also used for other seating positions and in different accident simulation tests, for example, for the rear seat occupant and for protection in side impacts. Detection of the accident severity in a side impact is much more difficult than in frontal collisions. And the time between the accident detection and deployment of the airbag system has to be very short, because in a 31 mph (50 km/h) movable barrier impact into the side of a vehicle, the occupant has already made contact with the interior side door panel after 25ms.

In the beginning, only airbags in the chest and pelvic areas were offered to protect occupants in lateral crashes. In the meantime, bags for side impact protection were expanded to protect the head and neck, as shown in Figure 6.13.

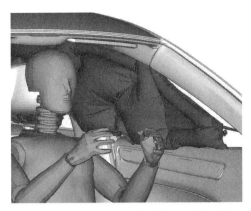

Figure 6.13: Side airbag, right side optimized for roll over protection [6-12].

The modified side airbag could also be used with the appropriate design to help to prevent the ejection of the head and other parts of the body in rollover accidents.

6.3.2.4 Seats, seat back, and head rests

Vehicle improvements that reduce risks from frontal and side collisions have focused more attention on rear-end collisions. Recovery from severe damage to the cervical spine often takes a long time. Therefore, new solutions to increase spinal protection became necessary. In one of the solutions, the head restraint is moved closer to the vehicle occupant's head in case of severe rear-end collisions.

The head restraint is moved by pyrotechnic, spring, or electric drives following activation by a rear-end collision sensor. Even the occupant's relative rearward movement to the seat back is used to protect the head. Some vehicles use small airbags. Figure 6.14 shows how Saab dealt with this problem.

6.4 Interaction of restraint system and vehicle

6.4.1 Unbelted occupant in a frontal collision

During a frontal collision of a vehicle with the impact speed v_i against a fixed barrier or another vehicle, the occupant moves because his forward inertia directs him toward the steering wheel or the dashboard and windshield. In a first approximation we find the following interrelationship:

Vehicle	Occupant until impact against vehicle interior
$\ddot{S}_V = -a$	$\ddot{S}_i = 0$
$\ddot{S}_{V(t)} = a \cdot t + v_i$	$\ddot{S}_i = v_i$
$S_{V(t)} = -\dfrac{a \cdot t^2}{2} + v_i \cdot t + s_o$	$S_i = v_i \cdot s_0$

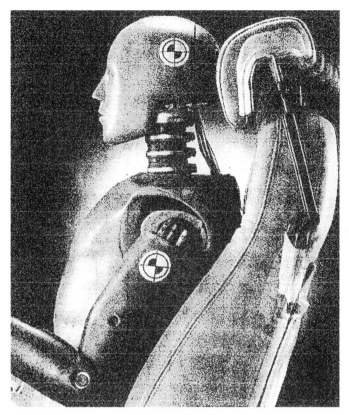

Figure 6.14: Headrest by Saab.

The relative movement between occupant and vehicle is as follows:

$$\Delta s = s_i - s_V = \frac{a \cdot t^2}{2} \tag{6.7}$$

Figure 6.15 shows the relative movement of the unbelted occupant and the crashed vehicle in principle.

During a 31 mph (50 km/h) impact against a fixed barrier with an average vehicle deceleration of 15 g and a distance of 11 inches (0.3 m) between the occupant and steering wheel before impact, the occupant hits the steering wheel after approximately 64ms. The differential speed is still approximately 21 mph (34 km/h). Without a restraint system the kinetic energy of the occupant has to be absorbed by the dashboard, steering wheel, and windshield. The deceleration level for the occupant of approximately 58g is higher than the 15 g deceleration of the vehicle. For comparison, the deceleration of a belted occupant is described in principle by the dotted line. The restraint results in lower maximum deceleration of the occupant.

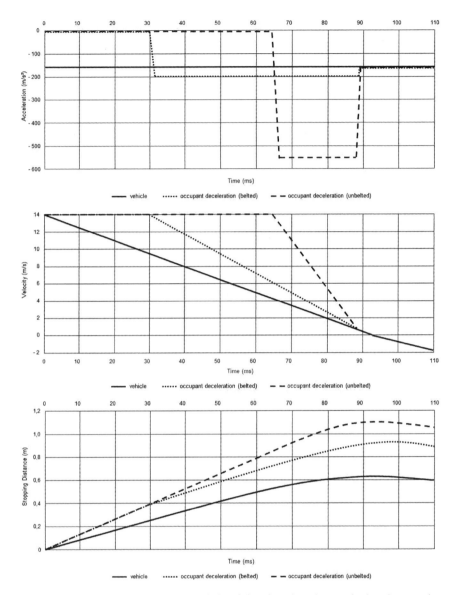

Figure 6.15: Principal characteristic of deceleration-time, velocity-time, and deformation time function of a 50 km/h crash test against a fixed barrier.

6.4.2 Belted occupant

When a safety belt without belt tensioner is used, the occupant still moves forward relative to the vehicle after the belt retractor locks and until the belt is tightened at the seatbelt reel or by the belt tensioner. The occupant is retained by the pelvic and torso belt. The pelvic belt exerts a vertical force downwards. For this reason the seat shell and its support have to be integrated into the total function of the restraint system. Figure 6.16 shows a typical deceleration time function for the head and chest and the belt forces as a function of time for a belted 50% male dummy in a frontal 31 mph (50 km/h) barrier crash.

To limit the relative forward movement of the occupants, especially for the driver, position belt tensioners are used. After a vehicle specific accident severity is reached, the pyrotechnic tensioner is activated via an electronic sensor. Because of the belt tensioner, the increase in the belt load starts earlier and the load level is lower compared to a standard three point belt. As a consequence the head rotation is also lower.

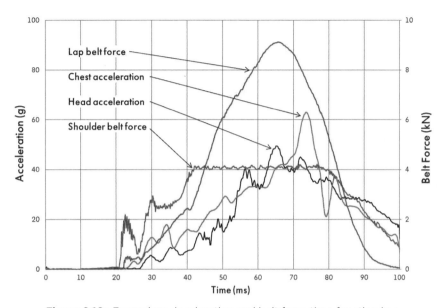

Figure 6.16: Exemplary deceleration and belt force time function in a 50 km/h barrier test with a 50% male dummy.

6.4.3 Airbag systems

The two different scenarios, airbag with the use of the three point belt and airbag without using the belt system, create difference occupant movement. In the first scenario the occupant moves like an occupant wearing only a three point belt in the beginning of the crash. After the airbag is inflated, the head rotation and head deceleration, as well as the chest load are significantly lower due to the additional support provided by the airbag system. In some tests the surface pressure of the dummy's head is measured with special equipment. The surface pressure at the head with an airbag steering wheel shows much lower values, compared to a steering wheel without an airbag. If no belt is used, the airbag system alone is responsible for the deceleration of the occupant. On the driver's side of the vehicle the airbag function is supported by the steering column and knee bolster and, in some vehicles by knee airbags.

On the passenger side, the airbag, knee bolster, and knee bag are responsible for deceleration of the occupant. Figure 6.17 shows a knee airbag on the driver's side of the vehicle. As the figure shows, the airbag reduces the occupant's contact with vehicle components like the steering column and steering wheel lock. In principle, an earlier contact with the vehicle is achieved; therefore, the stress values for the chest, pelvic area, and legs are reduced.

121

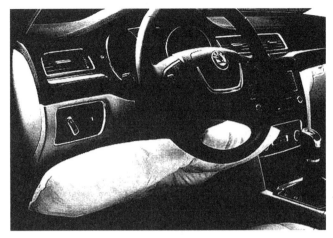

Figure 6.17: Knee airbag on the driver's side of the vehicle (source SKODA).

Situations where the occupant is not in a standard position, for example if he or she is not seated along the center line of the deploying airbag can be especially critical if the belt is not used. For this reason alone it is necessary that the three point belt is worn.

Today sensor and air bag hardware are common in most vehicles. Via software adaption the crash signal is analyzed and the first or second level of the pyrotechnic is ignited in accordance with the accident severity. The belt usage is checked. With the help of ultrasonic sensors, video, and seat mat or infrared sensors it is possible to monitor the occupancy of the passenger seat, an out of position situation, or the existence of a child restraint.

The situational airbag ignition is an adaptive design. The consideration of different occupants and accident severity is called adaptive occupant protection. Here different airbag rigidity and belt loads are possible.

6.4.4 Steering column deformation force

The steering column deformation can absorb additional energy. Current steering column deformation forces are approximately 3–5 KN. This means that the proportion of energy that can be converted is very low compared to the restraint system as shown in Figure 6.18. The advantage of deformation is focused on providing an additional distance for restraining the occupant. The deformation distance of current systems is about 2.4 to 3.9 inches (60–100 mm). Given nominal restraint distances of about 13.8 inches (350 mm) when the occupants are wearing seat belts, the steering column deformation can be converted into an additional energy absorption of approximately 25% by the seat belt and airbag restraint systems. Figure 6.18 shows the proportions of the energy that can be converted by a belt system, airbag system, and the steering column deformation at a 3 KN level. The calculation was performed using the restraint forces and deformation distance.

The force levels can be configured constant, with a degressive fall, or with a progressive increase. In principle, it can be advantageous to have different force levels

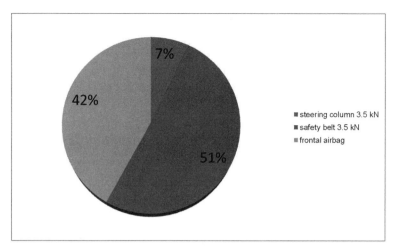

Figure 6.18: Proportions of the energy absorption of restraints composed of a seat belt, airbag, and steering column deformation.

for different occupant weights. A steering column force of 6 KN is significantly too high for the 5% female. In order to benefit from an additional restraint distances, the 5% female needs a level of about 3 KN.

Investigations have shown that an adaptive steering column deformation force is not required between the 5% female and 50% male if the adaptation of the airbag system is made variable by means of vent control. The reason for this can be found in the increased physical contact between the occupants and the airbag because the vent was initially closed. The occupant-dependent opening of the vent at different times can be used to achieve a comparable maximum airbag pressure (Figure 6.19, right). This produces an increased restraint effect and steering column force (Formula 6.8).

The rapid pressure increase in the airbag means that the lighter, smaller woman on the driver's side makes earlier physical contact with the airbag. At this time the vehicle deceleration results in an additional loading on the steering column. This is due to the mass moment of inertia of the steering column with the steering wheel and airbag (m_{SC} approx. 3.5 kg) depending on the crash pulse acting in the steering column direction (Figure 6.19, left). Consequently, the light 5% female also achieves a steering column deformation force of 3 KN over a longer period. The contact surface area of the 5% female is not much smaller than that of the 50% male, because the seat is positioned closer to the airbag so that the entire thorax makes contact over its height. In the 50% male, on the other hand, the airbag initially contacts only the upper thorax area because the seat position means that the pelvic area is still restrained by the belt.

Formula 6.9 applies with the airbag force acting on the occupant $F(t)_{OccupantAirbag}$ and the resulting steering column force F_{SC} according to formula 6.8 with the weight of the steering wheel including airbag m_{SC} and the crash pulse acting at the angle α in the steering column direction a_{CP}:

$$F_{LS}(t) = F(t)_{OccupantAirbag} + m_{LS} \cdot \cos \alpha \cdot a_{CP}(t) \tag{6.8}$$

$$F(t)_{OccupantAirbag} = A(t)_{OccupantAirbag} \cdot p(t)_{Airbag} \tag{6.9}$$

In this case, the restraint forces acting on the occupants differ according to the different contact surface areas $A(t)_{Occupant}$ of the occupants against the airbag during the crash (see also Figure 6.19).

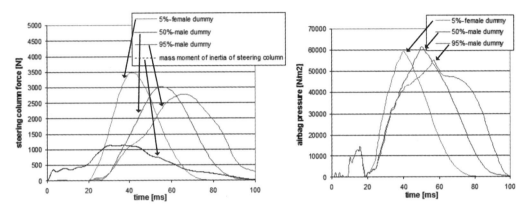

Figure 6.19: Steering column force (left figure) and rapid rise in the airbag pressure (right figure) due to the initially closed vent of the airbag control for 5% female dummy and the 50% and 95% male dummy. The mass moment of inertia of the steering column (SC) depends on the vehicle deceleration.

6.4.5 Optimizing the restraint system function

The belt system plays an important role in the total optimization of the restraint system. If we consider the effectiveness of the belt system in a real accident situation, it is significantly greater than that of the airbag system [6-13]. This is understandable because the belt system interacts with the vehicle at an early stage of the collision and can absorb very high restraint forces even without an airbag. On the other hand, the current design of the airbag reaches its limits much earlier when no belt is used, depending on the occupant weight and the severity of the accident. For example, there is a risk of puncturing from an accident severity with Δv of only approximately 19 mph (30 km/h). However, the disadvantages of the belt system lie in its small contact surface with the occupant and the associated high and locally concentrated loading on the thorax, as well as the lack of head protection. The airbag compensates for these limitations. Through interaction between the belt and airbag, it is also possible to reduce the belt loading on the thorax by using belt force limiters. This represents the current state of the art.

However, it is not possible to further reduce the belt force unilaterally for energy reasons. This can only be achieved sensibly in conjunction with the airbag control, in which case the airbag absorbs a higher proportion of the energy.

6.4.5.1 Concept of "less belt–more airbag"

The airbag control concept is particularly suitable for improved thorax protection because the belt force is reduced and, at the same time, the airbag force acting over a large area can be increased. The reduction in belt force must be exchanged for an increased energy conversion at the airbag. This is done by having a higher airbag pressure with the vent initially closed. To improve thoracic protection, the concentrated belt force acting on the thorax is reduced and the distributed force provided by the adaptive airbag compensates for it. The area of contact between the thorax and the adaptive airbag system amounts to $0.14m^2$ and is 3.9 times higher than the contact area of the safety belt system (Figure 6.20).

Figure 6.20: Thorax support over a large area by the use of an adaptive airbag system.

6.4.5.2 Ideal restraint effect

The restraint effect is the force which is applied to the occupant during the crash. The force influences the deceleration of the occupant significantly (force is related to occupant mass and acceleration: $F = m^*a$). The occupant restraint effect that can be implemented in a technical system differs significantly from the theoretical maximum possible effect. For example, the vehicle deceleration acting on the occupant (black profile in Figure 6.21) differs from the ideal square characteristic (red profile in Figure 6.21). If it were possible to connect the occupant directly to the vehicle then the ideal vehicle deceleration would follow the square characteristic. The occupant would then be decelerated in such a way as to be exposed to a minimum force level by using the maximum available forward displacement (travel includes the vehicle deformation travel and the free travel of the occupant in the vehicle interior).

Since the ideal vehicle deceleration cannot be implemented, the ideal occupant deceleration is provided in the red dotted profile in Figure 6.21. In this case, the occupant interacts directly with the vehicle and initially follows the vehicle's deceleration. As a result, the occupant is decelerated in such a way as to be exposed to a minimum force level by using the maximum available restraint distance.

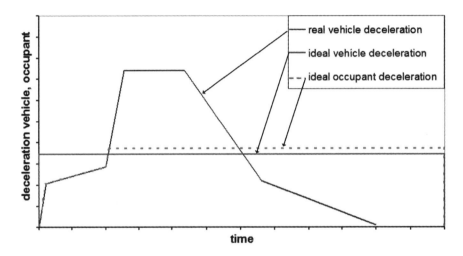

Figure 6.21: View (in principle) of the ideal (red profile) and real (black profile) vehicle deceleration and the resulting ideal occupant deceleration profile given direct interaction with the vehicle (dotted red profile).

The crash pulse is determined by the configuration of the structure. There are technical requirements which lead to different force levels of the front-end structure. The forces that can be converted in the area of the longitudinal member are rather low, to take into account collisions with pedestrians and minor obstacles. Involvement of the engine and the mechanical units, on the other hand, results in a significant increase in force and the maximum force level is defined by the cell rigidity (interior or survival space of the occupant). Precrash measures could optimize the crash pulse if active structures extend the structure (e.g., airbag in front of the vehicle) or the rigidity of the existing deformation length is increased [6-14].

Furthermore, the occupant cannot follow the vehicle deceleration immediately, because it is not technically feasible to achieve complete interaction with the vehicle by means of the safety belt system and the seat (see Figure 6.22, red curve). Active acceleration of the occupant in the opposite direction of the vehicle deceleration can, in principle, increase the level of interaction at the vehicle level. However, based on current knowledge, technical implementation of this does not appear to be practicable because of weight, cost, and functional reliability. The seat and the existing belt system cannot be infinitely rigid from a technical function perspective because of occupant comfort and the ability of the three point safety belt.

In principle, however, further optimization of the restraint function to achieve the maximum possible approximation to the square characteristic is possible. The forces that act on the occupant would then be lower in principle. The approximation can be achieved by a better interaction of the occupant with the vehicle and by limiting the force of the restraint system. The occupant deceleration that could be achieved and implemented in that case is similar to that of a trapezoidal characteristic. This is shown by the low and wide plateau formation in the green curve in Figure 6.22.

The tasks of the restraint system in achieving the plateau characteristic can thus be summarized as follows:

- Interacting early in the collision and participating in the vehicle deceleration
- Maintaining constant forces on the occupant as far as possible
- Exploiting the available forward displacement.

The question is what form the restraint system function must take in order to fulfill the aforementioned requirements to achieve the implementable trapezoidal characteristic.

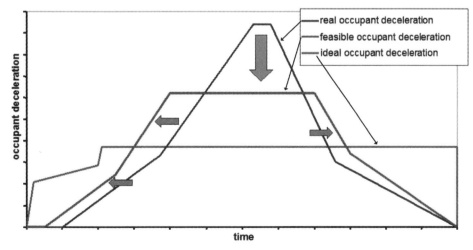

Figure 6.22: View (in principle) of the ideal occupant deceleration with initial direct interaction with the vehicle (red profile, see Figure. 6.21 red dotted profile) and the occupant deceleration that can be achieved by means of the trapezoidal characteristic (green profile) compared to the occupant deceleration of current systems (blue profile).

The functional profiles of an optimized restraint system for achieving the trapezoidal characteristic are calculated below using simulation methods on the basis of a mathematical dynamic model MADYMO with a 50% dummy according to the US-NCAP crash test. An iterative optimization was conducted to determine the optimum function profiles for the airbag and belt system. The relationship between crash pulse, restraint force, and occupant acceleration for the standard system and the optimized restraint effect is shown in Figure 6.23. For plateau formation (trapezoidal characteristic) in the occupant acceleration, it is necessary to have a restraint force that is declining degressively at the end of the crash. (Figure 6.23 left).

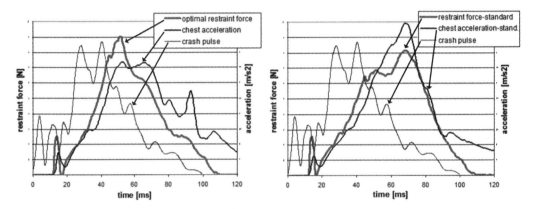

Figure 6.23: Crash pulse, restraint force (addition of the shoulder belt force and airbag force) and the resulting occupant acceleration for the standard system (right) and the optimized restraint function (left).

With the degressive restraint force, a high force at the beginning leads to early interaction and consequently participation in the vehicle deceleration already taking place. The occupant restraint effect (chest acceleration) is only initiated after a time lag, in contrast to the vehicle deceleration, due to the mass moment of inertia of the occupant, belt slack, and elasticity in the restraint system. The highest occupant loading occurs while the vehicle has already completed its maximum acceleration, which means that the majority of the energy has already been dissipated. This is then converted into a plateau characteristic by a degressive force reduction in the restraint system, in order to ensure that the forces acting on the occupant are as constant as possible.

The restraint forces on the occupant consist of the non-linear functions of the belt and airbag system (left Figure 6.24). These are available after different forward displacements by the occupant. In this case, the effect of the airbag comes into play after the effect of the belt. The belt system can be approximately represented by a two-stage force circuit (Figure 6.24 left). The degressive restraint effect of the airbag can be derived from the airbag pressure (Figure 6.24 right).

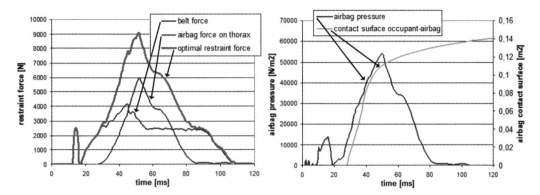

Figure 6.24: Addition of forces from belt and airbag force (left figure). The airbag force is given by the airbag pressure and the contact surface with the dummy (right figure). This increases with increasing interaction, so that the profile of the airbag pressure only corresponds to the airbag restraint force from the maximum pressure onwards.

6.4.5.3 Optimum profile of the belt force and airbag vent

To date, the optimization has involved establishing two belt force levels and one switching time for the belt force and determining one vent opening time and a corresponding vent opening diameter for the airbag vent control. For technical implementation, it is necessary to establish the ideal profile of these two parameters over time.

The optimum restraint system function should lead to a desirable approximation of the chest acceleration to the square characteristic. In principle, establishing the optimum function profile for the restraint system can be described by a physical replacement system [6-15], [6-16]. The disadvantage of the physical replacement system is, however, the associated simplification of the subsystems (deformable steering column, steering wheel, seat deformation, belt, and airbag system). By selecting the boundary conditions and system simplifications, different results are obtained [6-15], [6-16]. The system function of the belt and airbag is optimized by the evolution strategy. The advantage lies in the quality of the calculation, although the necessary calculation time is extensive. The two parameters to be investigated for the airbag and belt system are the belt force $F_B(t)$ and airbag vent diameter $A_{VD}(t)$.

To keep the calculation complexity within justifiable limits (9,000 individual calculations each taking about five minutes of calculation time, produces a computation run lasting 30 days), a time step width of 10ms was defined in advance. This means the function of the belt force $F_B(t)$ and the airbag vent diameter $A_{VD}(t)$ could only take on a new value every 10ms. The possible values for the belt forces for the belt retractors were also restricted between 1 KN and 6 KN (step width 10 N). Values between 0 inches (0 mm) (closed vent) and 3.1 inches (80 mm) (step width 0.039 inches (1 mm) were used for the diameter of the airbag vent.

The iteration of the search strategy was terminated after 30 generations, each with 300 populations after a maximum of 9,000 individual calculations. In order to ensure that the optimization of chest acceleration toward the required trapezoidal characteristic was not achieved at the expense of the two other target parameters, chest compression and head acceleration, the target parameter $PAIS3+$ was also selected here. The optimum restraint system function for the airbag vent diameter $A_{VD}(t)$ and the belt force $F_B(t)$ is shown in Figure 6.25 (left).

The numerically determined profiles show the necessary degressiveness for a rapid interaction between the occupant and the vehicle. The degressive force limitation can be seen both in the belt (drop in belt force in the left of Figure 6.25) and in the airbag (large vent opening requires initially closed vent with a degressive pressure drop in the airbag; see the left of Figure 6.25). Towards the end of the crash, the restraint forces of the belt and the airbag rise once again. In this case, the occupant's kinetic energy has been dissipated by the restraint system, and the occupant is at the transition phase towards return movement (rebound phase). The rise in restraint forces leads to an increase in the chest acceleration of the occupant. The occupant is already in the transition to the rebound phase, which means that no significant influence can be exerted on the forward displacement anymore. Accordingly, an increase in the restraint forces into the rebound phase is no longer essential with regard to energy.

The system function shown in Figure 6.25 for the airbag vent diameter $A_{VD}(t)$ and the belt force $F_B(t)$ is complex and cannot easily be achieved technically, because this would require variable belt switching and a vent permitting reversible opening and closing, with different vent diameter cross sections.

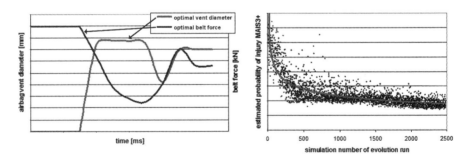

Figure 6.25: Left: Optimum restraint system function of the airbag vent diameter $A_{VD}(t)$ and the belt force $F_B(t)$ for the 50% dummy in the US-NCAP crash test Right: Optimization due to the Evolution Strategy; the probability of injury converges to a minimum.

A technically feasible solution combining a large vent that switched once and a switching belt force limiter would approach the optimum solution. The function profiles compared to the optimum solution are shown on the left of Figure 6.26. As explained above, there is no rise in the restraint forces during the rebound phase. The technically feasible solution displays significant potential for reducing the occupant loading values and leads to a desirable plateau characteristic for chest acceleration (right Figure 6.26).

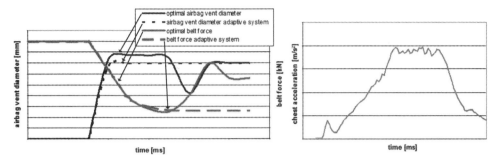

Figure 6.26: Left: Function of a single switchable airbag vent with safety belt load limiter (dotted line) compared to the idealized function of the airbag-venting and belt force. Right: Acceleration of the upper torso over the time for the idealized function of the adaptive system. Note the rectangular shape of the curve.

This technically feasible solution corresponds to a large extent to the concept of degressive vent control with a switching belt force limiter. The system correlation between a vent-controlled airbag and a switching belt force limiter is indicated below. In the first phase, during which the belt provides the only restraint, interaction is improved by the initially higher force level of the belt. During the phase in which the occupant is restrained by the airbag, the increasing airbag pressure results in an increasing total restraint force. The total force values would be excessive if the degressive airbag control were used in the case of maximum airbag pressure. The switching force limiter limits the total force in conjunction with the airbag force (Figure 6.27 right). However, the total force is also reduced during interaction with the airbag, resulting in a loss of interaction. The loss of interaction is made apparent by the drop in chest acceleration (Figure 6.27 right).

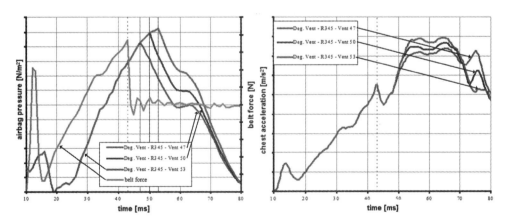

Figure 6.27: Airbag pressure and belt force, with the effect in chest acceleration. The addition of forces is shown by the variation in airbag pressure force in the acceleration diagram. In all three profiles, the belt force is identical and the only variation in the airbag is the time when the vent is opened by 3 ms.

The switching time of the belt force limiter can be varied relatively well during the transition of the occupant over to the airbag without negatively affecting the total restraint effect (maximum chest acceleration) (Figure 6.28). In conjunction with the degressive airbag vent control, it is consequently possible to implement a switching belt force limiter with relatively robust technical characteristics.

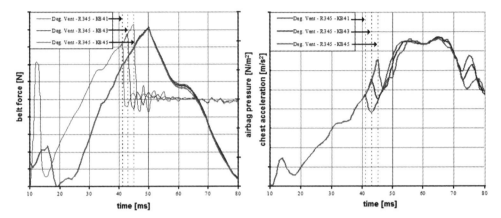

Figure 6.28: Robust configuration of the force limiter switching time: When the force levels are identical, the switching time can be varied without the total loading on the occupant rising (plateau level).

The brief but significant reduction in interaction with the occupant represents a disadvantage in the stepped switching of the belt force limiter, which is apparent by the profile of chest acceleration (Figure 6.28). A switching force limiter that can transition switching between high and low levels provides a remedy here (see Figure 6.29).

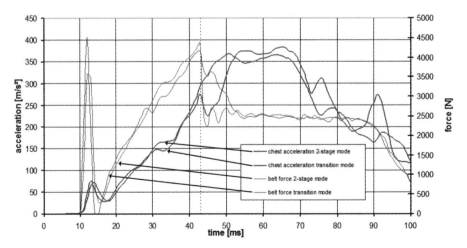

Figure 6.29: Seat belt (retractor) forces at the switching force limiter without (two-stage mode) and with transition function (transition mode) and effects on chest acceleration.

The effects of varying the second level of the belt force limiter with transition switching are shown in Figure 6.30. The magnitude of the belt force means that only the level of the restraint effect on the occupant is significantly influenced, not the functional profile. This is largely influenced by the degressive effect of the airbag.

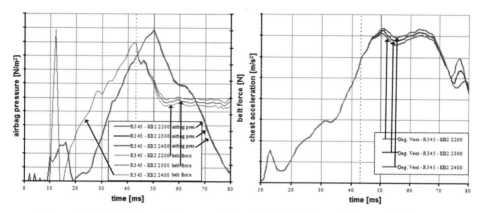

Figure 6.30: Variation of the second level of the belt force limiter and effect on the chest acceleration of the occupant.

Ideally, the belt force gradient of the transition switching of the belt force limiter is inversely proportional to the rising pressure gradient of the airbag. This would provide a homogenous transition between the belt force and airbag force acting on the occupant. Figure 6.31 shows the relationship between occupant deceleration and the forward movement of the occupant. The transition switching of the belt force limiter can consequently achieve a significant gain in interaction, which can be used either for a reduced forward movement or a lower total force level. The lower forward movement results in a greater safety reserve, while the reduction in the total force level reduces the occupant loading.

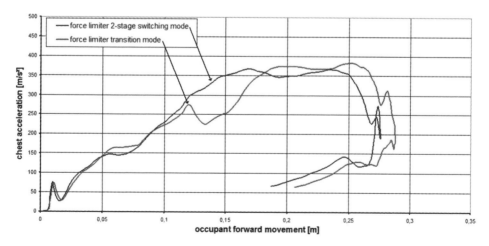

Figure 6.31: Chest acceleration/degree of forward movement characteristic of the occupant with vent control involving without (two-stage mode) and with transition switching of the belt force limiter.

The transition switching of the belt force limiter does reduce the loss of interaction, but at the same time produces an increase in chest compression. As a result, this is an irresolvable conflict of interest. The decision whether the force limiter should be switched with or without transition switching is consequently dependent on minimizing chest compression or chest acceleration. Weighting the relative changes tends to reveal a disadvantage for the transition switching. The force limiter with stepped switching reduces the total force during interaction with the airbag, which means that a loss of interaction is deliberately accepted in order to reduce the chest compression.

6.4.5.4 Steering column deformation force with a degressive airbag vent control

Steering column deformation provides additional restraint for the driver. Consequently, the movement of the steering column should be correlated with the movement of the occupant and his or her interaction with the restraint system. Excessively early movement would reduce interaction with the airbag, because the pressure build-up in the airbag would be reduced by "shifting away," reducing the counterforce.

Figure 6.32 shows this with reference to the flattening of the airbag pressure curve at 45 ms. The steering column deformation took place from 45–58 ms. After the vent opened at 50 ms, the airbag pressure initially decreased until the steering column deformation stopped. The travel limitation of the steering column results in a relative pressure rise. Figure 6.25 also shows the influence of the size of the vent opening on the restraint effect. Consequently, a rapid degressive drop in pressure would be favorable (blue profile). As the profile continues, an excessively rapid drop in pressure would cause the airbag to empty too quickly, which might lead to contact between the chest and the steering wheel (red profile).

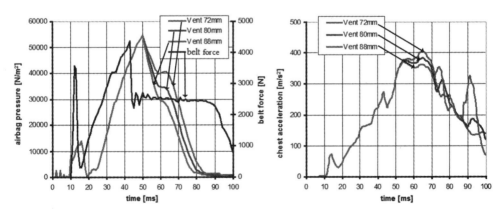

Figure 6.32: Influence on the pressure drop in the airbag by differently sized vent diameters (72 mm, 80 mm, and 88 mm), and the influence on chest acceleration.

In addition, correlation of the airbag with the steering column deformation provides an additional advantage. The degressive characteristic of the airbag pressure drop can be positively boosted if the deformation force of the steering column is adapted to the upper pressure level of the airbag by raising the break-loose moment). A favorable correlation would then be achieved by moving the steering column following interaction at the opening of the vent. Boosting the pressure drop at the same time as having the steering column move freely could reduce the vent diameter. This limits the pressure drop significantly towards the end of the crash (Figure 6.33), and improves the robustness of the system. This is also shown in the head acceleration (Figure 6.33).

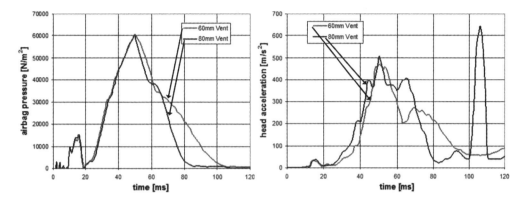

Figure 6.33: Airbag pressure and the resulting head acceleration for different vent diameter sand adjusted steering column deformation characteristics.

This correlation is also suitable for the 5% female dummy. Like the 50% male dummy, the steering column with the 5% female dummy can deform at a comparable pressure and force level of approximately 3 KN. The surface contact between the 5% female dummy and the airbag is lower, which means the applied force is correspondingly lower for the same airbag pressure. However, the lower force is compensated by the effect of the mass moment of inertia of the steering column/airbag unit.

6.4.5.5 Summary of optimization
The explanation of physical interrelations has shown that the restraint effect must be optimized using the trapezoidal characteristic approach. In this case, the trapezoidal characteristic represents the technical optimized approximation to the ideal square characteristic. Controlling the airbag vent diameter $A_{VD}(t)$ and belt force $F_B(t)$ appears promising in this case.

The concept includes the technical and physical requirements for reproducing the optimum restraint function, and thus leads to further optimized thorax protection and head protection. Airbag control with the vent initially closed leads to better interaction between the occupant and the vehicle deceleration. The subsequent degressive decline in airbag pressure through the opened vent limits the force on the occupant. The concept is particularly suitable for improved thorax protection, because the locally acting belt force is reduced at the same time as the airbag force, acting over a wide area, is increased. The concept also permits the required adaptive adjustment of the restraint effect by providing different airbag hardness according to the occupants and accident severities.

6.4.6 Lateral collisions

Because the free space between the occupant and the inner door panel is relatively small, the potential for an optimal restraint system for lateral collisions is much lower than for frontal collisions. The parameters that can help reduce the intrusion of the impacting vehicle are a collision-oriented design of the sill, the door beams, special reinforcements in the height of the shoulder, cross beams below the dashboard, reinforcements for the seats including the seat cross members, and crossbeams in the rear part of the vehicle. Other applicable parameters are the design of the vehicle interior, the shape and the material of the inner door panel, and the side trim. The airbag systems used to protect the occupant during side impacts depend very much on the functional sensor system which must be able to identify a severe accident early enough that the head of the occupant does not touch the side window.

6.4.6.1 Theoretical analysis

In comparison to the frontal impact, the number of variables in this type of accident is much higher including impact direction, first contact point, and geometric structures involved. In addition only small deformations zones are present. The influencing factors are as follows:

- Characteristics of the collision partner (mass, structure stiffness, structure geometry)
- Impact point and angle, impact speed, and occupant seating position
- Design of the vehicle interior
- Use of the restraint system

In Figure 6.34 the vehicle to vehicle side crash is explained in principle. The impacting and impacted vehicles with one occupant on the impacted sides are shown. Also the figure illustrates one example of a simple mathematical model for the occupant and for the vehicles as a velocity-time function. The numbers 1 to 7 identify the velocity time function of the vehicle components and the occupant. The narrowed areas correspond to the relative movement (deformation) between the vehicle component and between the components and the occupant.

The force-displacement characteristics for the structure and inner side panel are defined as rectangular. Oscillations of the mechanical mathematical model are not considered. Therefore the calculated functions $a = f(t)$ and $s = f(t)$ differ from a real world lateral collision. Because of this limitation and the single mass occupant model, only a qualitative comparison in the following discussed collisions mode is possible.

The impacting vehicle drives with a velocity v_{St} under 90° into the side of the impacted vehicle ($v = 0$).

After a short time the outside door 5 reaches the same velocity as the impacting bumper 6. The inner door 3 has, at t_T, the same velocity as the bumper. The contact between occupant and restraint system 3 starts at time t_T. At t_T the relative velocity between occupant and door is highest. It is higher than the change of velocity of

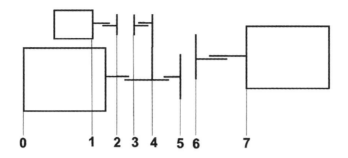

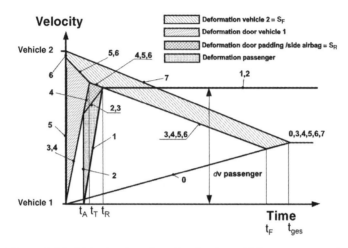

Figure 6.34: Velocity-time-function for the vehicle structure and occupant in a right angle lateral collision.

the impacted vehicle. The occupant is decelerated by the bolster. The deformation is S_R. Occupant 1 and bumper 6 have, at t_R the same velocity. Until this the occupant and the bumper are decelerated equally. The deformation of the side structure is finished at t_F. That means no further deformation is possible. The front structure of the impacting vehicle was deformed by s_F. The deformation of the side structure is defined by the relative movement of door 3 to an undeformed point of the impacted vehicle.

6.4.6.2 Side impact test defined in the U.S. and Europe

For the simulation of dynamic vehicle to vehicle collisions the new FMVSS214 is being introduced gradually (in production cars from September 1, 2009 to September 1, 2013). The conditions for the new 214 are as follows.

Barrier: 1.368 kg mass, width 1.676 mm, ground clearance 279 mm

Impact speed: in crabbed direction 54 km/h (48 km/h in 90° direction)

Test dummies: ES-2re front, SID-IIs rear at the impact side

Criteria: S10 IIs: HIC 36 < 1000 [–] lower spine acceleration
< 82 [g] pelvic force <5.525 KN

              ES-2re: HIC 36 < 1000
chest deformation < 44 mm
abdomen force < 2,5 KN
PSPF < 6 KN

For Europe the requirements ECE-R95 and 96/27 EC are valid:

Barrier: 950 kg mass, width 1.500 mm, height of the barrier front
800 [mm] ground clearance 300mm

Impact speed: 50 km/h

Test dummies: ES-2 front

Criteria: MPC <1.000 [–], viscous criteria < 1m/sec chest deformation
< 42mm, abdomen force < 2,5 KN, PSPF < 6KN

For the NCAP test in the United States the barrier and test dummies used in the new FMVSS 214 are evaluated but with a higher speed of 37.9 mph (61 km/h) (55 km/h in 90° direction). For the NCAP evaluation several countries have defined different test and evaluation criteria. Countries like Japan, China, Korea, and Australia rely more on the ECE-Test. The new pole-side impact test for the NCAP evaluation also includes differences in the impact direction and speed as well as requirements for the dummies and the criteria for passing the tests.

These differences in the legislation also have a negative impact on the vehicle design. In the beginning of the development of safety measures for lateral impacts the vehicle itself, the resistant body in white, and energy absorbing vehicle interior, were the basis for occupant protection. After the installation of airbags in the seat back or door panel, more and more torso and head airbags are being built into the vehicles. In the meantime, due to the requirements of the containment test in the U.S., the design of the side head airbag has to be improved.

In any case the performance of the side airbag must meet higher standards. Due to the shorter distance of the occupant from the vehicle's side, the sensor and airbag ignition and inflation have to work much faster than they do in a frontal impact.

6.5   References

6-1   FMVSS 210, Seat belt assembly anchorages Federal Motor Vehicle Safety Standards, NHTSA USA.

6-2   FMVSS 216, Roof crush resistance, Federal Motor Vehicle Safety Standards, NHTSA USA.

6-3   FMVSS 214, Side Impact Protection, Federal Motor Vehicle Safety Standards, NHTSA USA.

6-4 Wittemann, W. Improved Vehicle Crash Worthiness Design by Control of the Energy Absorption for Different Collision Situations. Technical University of Eindhoven, The Netherlands 1999, ISBN-90-386-0880-2.

6-5 FMVSSS 208, Occupant crash protection, NHTSA, USA.

6-6 Carhs, Safety Companion 2012 Carhs Training GmbH, Alzenau Germany.

6-7 FMVSS 301, Fuel System Integrity, Federal Motor Vehicle Safety Standards, NHTSA USA.

6-8 FMVSS 226, Ejection Mitigation, Federal Motor Vehicle Safety Standards NHTSA USA.

6-9 http://www.seat.de/seat_/magazin/technik-lexikon/p/pyrotechnischegurt straffer/contentParagraphs/0/image/pyrotgurt.jpg (October 2010).

6-10 ECE R 44.04, ECE R 44.03, Uniform Provisions Concerning the Approved of Restraining Devices for Child Occupant of Power Driven Vehicles. United Nations, Geneva, Switzerland.

6-11 http://www.volkswagen-zubehoer.de/shop Kindersitz.html (Oktober 2010).

6-12 Borgmann, Ph. Sicherheit bei Fahrzeugüberschlägen für nicht angegurtete Insassen. Technical University Braunschweig, 2012, Verlag Dr. Hut, München.

6-13 Zobel, R. Benefit from Fleet Change and Restraint Systems, 9. EMN Congress Bucharest, Romania, May 20–24, 2004.

6-14 Shi, Y., Wu, J., Nusholtz, G.S. Optimal frontal vehicle crash pulse—A numerical method for design, ESV Paper No. 514, 2003.

6-15 Hesseling, R. Active Restraint Systems–Feedback Control of Occupant Motion, Dissertation, TechnischeUniversität Eindhoven, Netherlands, 2004.

6-16 Ikels, K. Innovative Energiemanagement Methode zur Analyse und Verbesserung von Insassenschutzsysteme, EASi Engineering GmbH, München, 2001.

Chapter 7
Adaptive Occupant Protection

7.1   Requirements based on the accident situation

An analysis of the accident situation makes it possible to identify the most important accident scenarios in which an adaptive restraint system could prove effective. Summarizing the individual accident scenarios reveals the target population. Starting from this target population, it is possible to determine occupant and accident-specific influencing parameters, the significance of which describe the adaptively requirements that are possible and can be implemented.

An adaptive restraint system should be triggered variably depending on the accident severity and the individual properties of the occupants. Ideally, this applies to all accidents and all occupants. However, the adaptive restraint system cannot be implemented equally for each accident. This is because of the physical requirement of the adaptive system which is only able to operate optimally by completely exploiting the available restraint distance with minimum restraint force. In this case, the level of the restraint force is principally determined by the occupant and the severity of the accident. If the forward movement that can be used to decelerate the occupant is restricted, because of failure to use a seat belt, intrusion of an object or vehicle into the passenger compartment, or the type of collision (e.g., rollover), then the potential for an adaptive restraint system is low. In those accidents in which an adaptive restraint system can prove effective, the target population is described below, based on the accident database. This assumes that an adaptive restraint system can offer additional protective potential for all occupants and accident severities within this target population. The following analysis of real-world accidents is based on the GIDAS accident database.

The accident scenarios can be divided into various types of collisions. The frontal collision is not just the most frequent type of collision, but frontal collisions also cause the most serious injuries (MAIS3+) and fatalities (MAIS5+: includes MAIS5 and MAIS6 injuries). An adaptive system could prove particularly effective here because the available restraint distance permits variable control of the occupant.

In contrast, the restraint distances available in side, rear, and rollover collisions are significantly shorter; consequently the effectiveness of adaptive systems in these collisions tends to be rather limited.

In passenger car collisions, the collision participant has a decisive influence on the risk of injury. Evaluation of frontal collisions shows that the most frequent MAIS3+ injuries are recorded in car-on-car collisions, collisions with trees and pillars, and with heavy and light commercial vehicles (Figure 7.1). In car-on-car collisions, the significant difference is between high frequency and lower risk. This includes, above all, many minor collisions which reduce the risk. In collisions with trees and pillars, the collisions are frequently ones involving a higher Δv. This means both the frequency and the risk are high.

For car occupants, the effect of an adaptive restraint system in collisions involving heavy commercial vehicles tends to be low. In such cases, an under-ride guard on the truck would be more effective. Due to the significant difference in mass and the lack of geometrical compatibility between trucks and cars, the occupant compartment of the car is frequently overloaded, and the truck intrudes into the compartment. The intrusion, deformation, or penetration of components into the passenger compartment reduces the survival space of the occupants, and consequently the available space for operation of restraint systems. A functional, adaptively operating restraint system requires a prediction function for the expected intrusion. Without this prediction, the degree of forward movement of the occupants would be incorrectly calculated. The restraint effect would then be insufficient, and the occupant would impact against parts of the interior. However, predicting the intrusion is currently not feasible. Too many parameters influence the structural deformation, and thereby, the possible intrusion in the various accident scenarios.

The analysis of real-world accidents based on the GIDAS accident database reveals a target population for adaptive restraint systems in frontal collisions. This target population for adaptive restraint systems can be summarized as 31% of all car

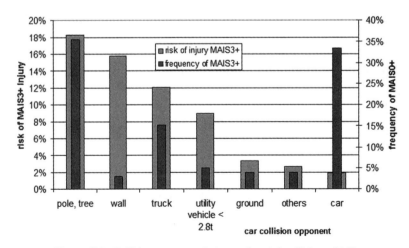

Figure 7.1: Collision opponents in car frontal collisions [7-1].

occupants with MAIS3+ injuries with restrictions made according to the type of collision, collision participants, required seat belt usage, and the requirement for cell integrity (Figure 7.2).

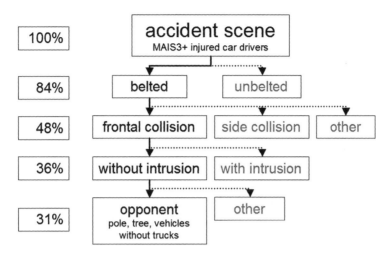

Figure 7.2: Target populations of adaptive restraint systems [7-2].

7.2 Individual occupant protection

Within the aforementioned target population, an adaptive restraint system should improve occupant protection. Issues that affect individual occupant protection include specific accident and occupant conditions.

7.2.1 Accident severity

Accident severity has a decisive influence on the risk of injury to the occupants (Figure 7.3). While a restraint system may be too harsh for less severe accidents, in accidents that are significantly more severe than the design speed the restraint system may be overloaded.

The accident severity is described by the variable Δv. The value Δv describes the change in speed of the vehicle during the collision and can be used to provide a good estimate of the actual accident severity.

It is obvious that the risk of suffering a severe injury (MAIS3+) increases with increasing accident severity. The question—in which accident severity range should an adaptive restraint system take effect—can be answered by weighting the risk over the frequency of accidents in the individual accident severity classes.

The weighting is performed on the basis of the MAIS3+ injuries according to the following approach:

$$r_{norm_{MAIS3+}}(\ v_{Class}) = \frac{h_{MAIS3+}(\Delta v_{Class}) \cdot r_{MAIS3+}(\Delta v_{Class})}{r_{mMAIS3+}(all\_\Delta v_{Class})} \tag{7.1}$$

143

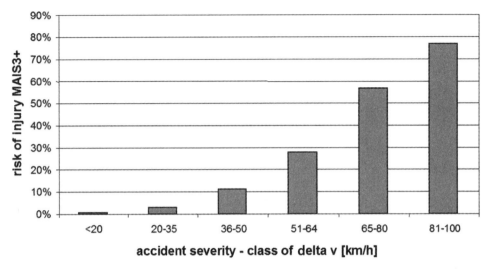

Figure 7.3: The injury risk of the occupant is largely determined by the accident severity.

The formula includes the Δv class-related frequency $h_{MAIN3+(\Delta vclass)}$, the Δv class-related risk $r_{MAIS3+}(\Delta v_{class})$ and the medium risk $r_{mMAIS3+}(all\_\Delta v_{classes})$. Figure 7.4 shows the weighted and normalised risk $r_{normMAIS3+}$ over the accident severity classes Δv. The weighting indicates an accident severity range of 22.3 to 23.6 mph (36–80 km/h). Consequently, an adaptive restraint system could achieve its greatest potential within this range. The risk of injury is already very low at lower accident severities, consequently adaptivity does not appear to be advantageous in spite of the greater frequency of these accidents. On the other hand, the risk of injury is high in very high accident severities, but the frequency is very low. In addition, it seems

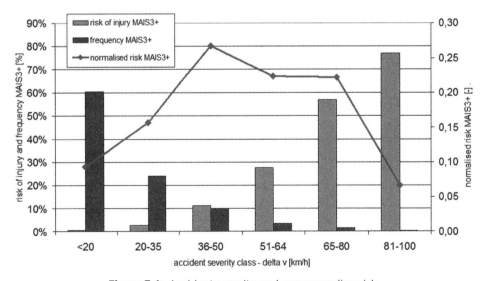

Figure 7.4: Accident severity and corresponding risk.

questionable to what extent passive safety measures have anything to offer at very high accident severities. Here, the overriding goal should be to reduce the accident severity in the course of developing active safety. As a result, it would be better to address accidents with particularly high accident severities by combining active and passive safety.

The deceleration time curve of the vehicle is designated the crash pulse. The crash pulse depends not only on the collision direction and impact speed, but also on the body structure and mass ratio of the vehicle and the other vehicle involved in the collision. In terms of determining the occupant loading, the crash pulse represents a significant building block in simulating a real accident.

During an accident, kinetic energy must be dissipated, which means the speed change Δv undergone by a vehicle in a collision represents a measure of the severity of an accident. Simultaneously, the time during which the kinetic energy is converted also plays a role, because it determines the acceleration level of the the occupants. The magnitude and duration of occupant deceleration significantly determines the loading on the occupant.

Deceleration time profiles measured on the vehicle body are subject to significant superimposed noise. In order to eliminate high-frequency components, these can be smoothed out subsequently using standardised filters. A filtered deceleration profile of this kind is shown as the black curve in Figure 7.5.

The deceleration profile can be built up as an approximation from linear functions to describe the characteristic procedures in the crash (see Figure 7.5, red profile). The first linear rise describes the deformation of the bumper and the deformation

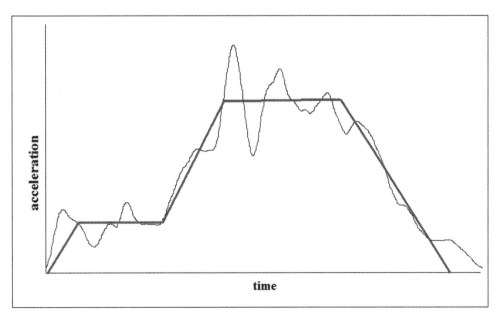

Figure 7.5: Example of a measured deceleration profile and the approximation by linear functions (red profile).

box directly after impact with the obstacle. The following section, with constant deceleration, represents the free deformation length of the front end of the vehicle. The linear rise that follows is due to the contact between the vehicle's engine and the bulkhead wall. This is associated with an increase in rigidity of the vehicle, and consequently, an increase in deceleration. The subsequent area of constant decelerations represents the elastic and plastic deformation of the remaining part of the front end. After the kinetic energy of the vehicle has been dissipated, deceleration declines.

The deceleration profile can be built up as an approximation from linear functions in order to describe the characteristic procedures in the crash (see Figure 7.5, red profile). The first linear rise describes the deformation of the bumper and the deformation box directly after impact with the obstacle. The following section, with constant deceleration, represents the free deformation length of the front end of the vehicle. The linear rise that follows is due to the contact between the vehicle's engine and the bulkhead wall. This is associated with an increase in rigidity of the vehicle, and consequently, an increase in deceleration. The subsequent area of constant decelerations represents the elastic and plastic deformation of the remaining part of the front end. After the kinetic energy of the vehicle has been dissipated, deceleration declines.

The decelerations measured on the vehicle do not apply to the occupant in this way; however, because the occupant is not rigidly connected to the vehicle, but is only linked to it by means of the restraint system and the seat. As a result, the measured crash pulse is filtered from the perspective of the occupant. For the purposes of the investigation being performed, it is necessary to use crash pulses which cover not only the greatest possible range of accident severities but which can consider the accident events adequately. Different accident severities and accident events result from the analysis of accident research.

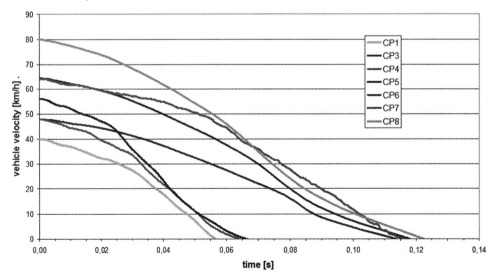

Figure 7.6: Variation in some of the crash pulses used in the speed/time representation (CP1: FMVSS208-25mph, CP4:FMVSS208-30mph, CP5: USNCAP, CP7: EUNCAP).

Statistical evaluations of accident research show that in 60% of cases when the accident involves a frontal collision with a speed difference of more than 20 km/h, the other vehicle involved in an accident is a passenger car. The OBD test according to Euro NCAP is very well suited to represent vehicle/vehicle collisions [7-3]. Furthermore, the crash pulses in the most important registration tests according. to FMVSS-208 and US-NCAP are used. These also reproduce a vehicle/vehicle test in part [7-4] (for the U.S. market, about 74% of frontal collisions can be effectively covered by the barrier test in FMVSS 208), but result in a more severe loading on the occupants due to the significantly greater deceleration values and shorter crash pulse durations. The more meaningful speed/time profiles can be shown by integrating the deceleration time profiles or crash pulses.

7.2.2 Individuality of the occupants

Dummies are used in crash tests to achieve a quantitative description of the occupant loading. These are anthropometric measurement dummies which simulate the weight, mass distribution, and movement possibilities of the human body. The physique of vehicle occupants varies widely, consequently the so-called 50% dummy is used. This represents the average occupant. In order to take account of additional vehicle occupants who vary widely from this average value in terms of size and weight, there is a range of additional dummies. The relevant ones are the 95% male and the 5% female. Only five percent of men are taller than the 95% man, and only five percent of women are smaller than the 5% female. This means a wide subset of the actual population can be represented. Figure 7.7 shows the three dummies.

Sensors are attached to the head, chest, hips, and thighs of the dummies to calculate the protection criteria by measuring the accelerations, compressions, and forces.

These three hybrid III dummies were used in the model. As can be seen from the accident analysis, these largely correspond to the real occupants. The total frequency distribution of occupant weight and size derived from the total accident situation is shown in Figure 7.8. The 50 percentile with a weight of 160.9 pounds (73 kg) and an occupant height of 68.5 inches (174 cm) corresponds approximately to the 50% male dummy. The 5 percentile with a weight of 114.6 pounds (52 kg) and an occupant height of 62.2 inches (158 cm) corresponds approximately to the 5% female dummy. The 95 percentile with a weight of 216 pounds (98 kg) and an occupant

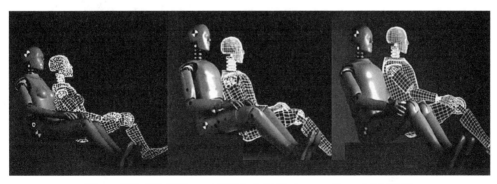

Figure 7.7: 5%, 50%, and 95% hybrid III dummy (from left) [7-5].

height of 73.6 inches (187 cm) corresponds approximately to the 95% male dummy. The slight deviations can be explained by the influence of sex and would be significantly lower if the consideration in Figure 7.8 were sex-specific. (The information in Figure 7.8 contains both female and male car occupants involved in accidents, in a ratio of about 2:3. The sex-specific representation was not differentiated in this case.) Consequently, the occupant size/weight distribution in the investigated accident data correlates to the crash test dummies used and gives a representative impression of the total distribution.

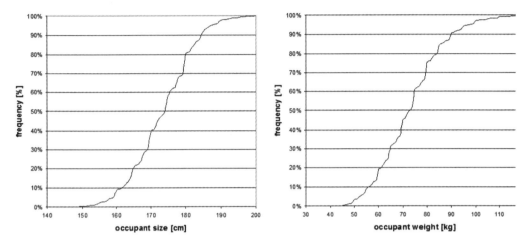

Figure 7.8: Total frequency of occupant size and occupant weight in the accident data. The 5, 50 and 95 percentiles correspond to the 5%, 50%, and 95% dummies.

The accident risk for the different occupant sizes and occupant weights is determined below in accordance with the target population of car drivers wearing seat belts in frontal collisions. It can be seen that there are different injury risks between the occupants, (see also [7-4]). The influence of the different occupant sizes on the *MAIS3+* injury risk is shown in Figure 7.9.

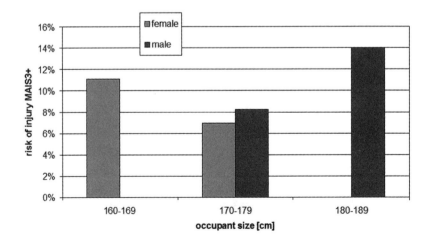

Figure 7.9: Change in the MAIS3+ injury risk based on occupant size and according to sex.

Starting from an optimum value, corresponding to the 50 percentile size, the injury risk increases both with increasing and declining occupant sizes. It should be noted that there are non-resolvable interactions between occupant size, sex, weight, and seat position. As far as the driver is concerned, for example, his or her size, and consequently also sex, has a significant influence on the seat position. The increase in injury risk for small vehicle occupants seen in Figure 7.9, can be explained by the interaction between the occupant's sex and the seat position. This includes a large proportion of women in the front seat. However on average, as the occupant's size increases so does his or her weight.

In principle, the probability of suffering a severe injury is greater for heavier occupants (refer to the black second order polynomial trend line in Figure 7.10), because the restraint system has to absorb more energy if a greater weight is involved. In contrast, the restraint system can tend to be excessively harsh for lighter occupants. In this case, the available restraint distance would be inadequately used.

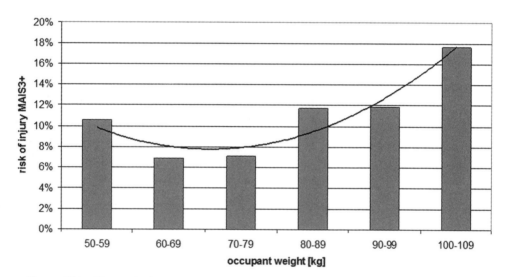

Figure 7.10: Change in the MAIS3+ injury risk over the occupant rate with a second order polynomial trend line (black profile).

7.2.2.1 Seat settings

The seat position of the vehicle occupants plays an important role in an accident. It defines the maximum available restraint distance and influences the kinematics of the occupant. The seat position can be described by three parameters: backrest angle, seat forward/backward movement, and seat height. The variability of the seat position results in different distances from the steering wheel (Figure 7.11) [7-5].

Seat position correlates in particular to the occupant size. The distances of small women (5% women up to a height of 60 inches [153 cm]) from the airbag module were on average 12 inches (308 mm), which is adequate. This value is of particular

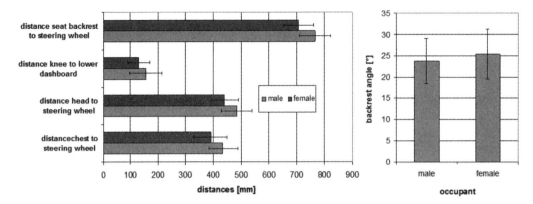

Figure 7.11: Average values and standard deviation of the distances of the occupant from the steering wheel and dashboard, and backrest angle of the driver seat.

interest for an adaptive airbag size. Very close distances of less than 10 inches (254 mm), are likely to result in airbags that are out of position, as observed by [7-6]. However, this situation was seen in only 0.3% of the cases.

A variation in the seat backrest angle may change the occupant's kinematics. For example, if the backrest is tilted back, the occupant has a tendency to submarine or slide under the seat belt while there is an increased likelihood of upper body rotation if the backrest angle is steeper. The variation in possible backrest angles was measured in the seat study. According to this study, the statistical mean of the backrest angle corresponds to the backrest angle set in the simulation and in the crash test at approximately 23–25° with a standard deviation of 5° (Figure 7.11).

In the seat study, it was possible to observe a clear dependency between seat height and occupant size. Small drivers tend to choose a higher seat position, while tall drivers choose the lower one. The accident analysis did not reveal any significant influence with regard to seat height, because the data were not sufficiently detailed.

7.2.2.2 Age of the occupants
The injury risk increases with the increasing age of the occupants (Figure 7.12) because of the reduction in biomechanical resistance of the body. Studies [7-7] [7-8] show that bone strength declines starting at 35 years of age, and is only about 45–75% in a 60 year-old person.

Figure 7.12 also shows an increase in the injury risk for younger occupants because of an interaction between the accident severity and the age of the vehicle in which they are traveling. The higher accident severity is an expression of the increased risk acceptance level and lack of driving experience of younger people. For financial reasons, young drivers also tend to drive somewhat older vehicles, which consequently might not reflect the latest safety standards.

Considering the decline in biomechanical loading capability of older occupants, the question arises as to what extent age-dependent occupant protection can be implemented at all. [7-9] Based on accident investigations, a 2 KN belt force level

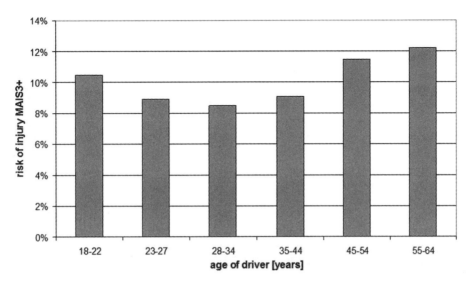

Figure 7.12: Influence of occupant age on the MAIS3+ injury risk.

with seat belt force limiters would be advantageous for a 70 year-old person to effectively reduce the risk of fractured ribs. However, reducing the belt force on one side reduces the restraint effect. In terms of energy, increasing the airbag force acting over a large area can compensate for the reduction in the locally acting belt force. In this case, the airbag has to absorb a higher proportion of the energy. The loading distribution over a larger surface area could reduce the thorax loading accordingly.

Minimizing thorax loading is especially useful for older occupants. Although younger occupants can withstand higher loadings, they are also subject to a correspondingly reduced thorax loading given the same accident severity. If the configuration of the restraint system is adapted to different accident severities, then age-dependent occupant protection is no longer necessary because the magnitude of the restraint force can be minimised irrespective of the age of the occupant. Additional protective measures for older occupants and using sensors to determine an occupant's age are consequently not required.

7.2.3   Weighting of the main influencing factors

By means of normalization, and with regard to the same general set of data material, the occupant and accident-specific main influencing factors (MEFs) described above can be weighed off against one another. To do this, the MAIS3+ risk of the individual main influencing factors $r_{MAIS3+}(MEFs)$ is determined. In this case, the risk of the main influencing factors is normalized $r_{normMAIS3+}(MEF)$ to the MAIS3+ risk $r_{MAIS3+}(target\ population)$ of all accidents within the aforementioned target population, according to equation 7.2.

$$r_{normMAIS3+}(HE) = \frac{r_{MAIS3+}(HE)}{r_{MAIS3+}(Active-field)} \qquad (7.2)$$

Figure 7.13 shows the weighting for the influencing factors as well as the corresponding setting for factors for the high and low injury risk in the Δv range of 21.7 to 49.7 mph (35–80 km/h). It would be desirable for an adaptive restraint system to deliver additional protective potential for different occupants within this range.

The weighting of the individual influencing factors is determined by the deviations in the injury risk in relation to the averaged risk. In this case, it is assumed that if there is an increasing injury risk, then unfavorable boundary conditions apply, which might be positively changed by optimizing and using an adaptive restraint system.

The weighting makes it possible to derive the factors that have a decisive influence on the risk of injury, and which consequently, should be taken into account by an adaptive restraint system. These are described here mainly by the occupant weight and the accident severity (Figure 7.13). The accident severity, the change in velocity of the vehicle during the collision, and the occupant weight describes the kinetic energy of the occupants that has to be absorbed by the restraint system. Consequently, it would appear promising for the restraint effect to be adapted to the particular accident severity and occupant weight.

In terms of the configuration of adaptive restraint systems, the useful forward displacement of the occupant, and consequently the seat front/back position, also appear to be of interest, because this determines the magnitude of the restraint force. However, additional sensors are not required for the driver because his or her weight, size, and seat front/back position largely correlate with one another.

The occupant age is also an important parameter. However, this does not have to be considered during development of adaptive systems, because the accident severity-specific adaptively makes age-dependent protection superfluous.

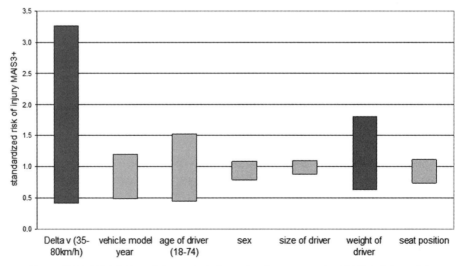

Figure 7.13: Weighting of the main influencing factors on the MAIS3+ injury risk. The values in brackets specify the range of main influencing factors considered.

7.3 Airbag control concepts

Former considerations show that a restraint system to be designed with adaptive characteristics must be significantly shown by the airbag and belt system.

The previous chapter determined the principle profile of the airbag and belt force function. The different concepts for technical implementation are evaluated below. In this case, the degressive airbag force is evaluated over the time profile of the airbag pressure $p_{Airbag}(t)$ and the degressive belt force over the time profile of the belt force $F_B(t)$ in the shoulder belt.

The hardness of the airbag can be described by the time profile of the airbag pressure $p_{Airbag}(t)$. The equation of states of ideal gases can be used for estimating the influences on the airbag pressure (Formula 7.3). The simplified assumption applies that the gas mixture in the airbag approximately behaves like an ideal gas:

$$p_{Airbag}(t) = \frac{v_{Airbag}(t) \cdot R \cdot T}{V_{Airbag}(t)}. \tag{7.3}$$

Formula 7.3 shows that the airbag pressure can be influenced by the airbag volume V_{Airbag} and the amount of substance V_{Airbag} in the airbag. The amount of substance in this case is directly dependent on the mass of gas m_{Airbag} (with the molecular mass $M: v_{Airbag} = m_{Airbag}/M$ [g/(g/mol)]). Parameters which cannot be adaptively controlled are the gas temperature T and the general gas constant $R = 8.314 Pa \cdot m^3/mol \cdot K$.

The airbag and belt system can consequently be varied in their hardness by basically four different control measures:

- Varying the timing of supplying the mass of gas for filling the airbag: $m_{Airbag}(t)$
- Varying the timing of the airbag volume $V_{Airbag}(t)$
- Varying the gas outlet by variable blow-off openings (vents): $A_V(t)$
- Varying the belt force over time: $F_G(t)$

7.3.1 Mass flow control

By controlling the quantity of gas produced from the gas generator, it is possible to vary the amount of gas in the airbag, and consequently the airbag pressure. The gas is generated by combusting a solid fuel, or alternatively, by opening a pressure accumulator (cold gas generator) which stores a compressed gas. Technical limits are imposed on the continuously steady gas generation and limit the variability of the mass flow profile by scaling the mass flow. Tailoring the propellant makes this technically feasible.

The mass flow of the validated airbag system was varied by a scaling between 60% (mass flow factor 0.6) and 127% (mass flow factor 1.27). Figure 7.14 shows the corresponding airbag pressure profiles. These result from the unfolding of the airbag, the filling of the airbag, and subsequent immersion of the occupant in the airbag. Consequently, the mass flow increase results in an increase in the airbag

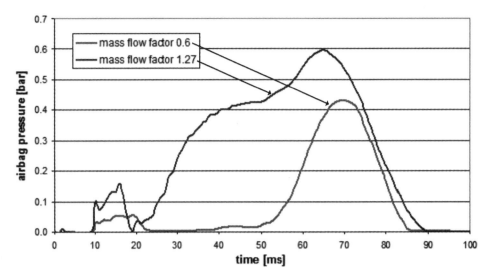

Figure 7.14: Airbag pressure profile by scaling the mass flow factor.

pressure in the initial period. This leads to an earlier and stronger interaction with the occupant, which means the interaction is more effective. Associated with this, however, is a significant increase in the maximum internal pressure, leading to an overload on the occupant. This overload occurs because the gas cannot be channelled out of the airbag quickly enough. This is a typical conflict of interest in the mass flow control.

To limit the airbag pressure and consequently limit the force on the occupant, it is possible to use a comparatively larger vent. In that case, the gas can be dissipated more rapidly from the airbag with a reduced choke point resistance. In order to compensate for losses at the beginning, however, the mass flow must be increased significantly. By way of comparison, Figure 7.15 shows a mass flow profile (violet dashed curve) which produces the required degressive airbag pressure (violet curve). From a technical perspective, the ability to implement this mass flow profile is only possible to a limited extent because the gas is not generated in a continuous, steady stream, and an elaborate control system is required for the input mass flow.

Variable mass flow profiles can also be achieved by two-stage gas generators. Following ignition of the first stage of the generator, time-delayed ignition is used for the second gas generator stage. However, the variability of the mass flow is restricted because the amount of propellant in the two stages must be defined in advance. The mass flow profile that can be implemented in a two-stage gas generator system is shown in Figure 7.16 for different ignition times and gas masses used in both stages. The necessary mass flow (see Figure 7.15, violet dashed curve) deviates significantly from that which can be implemented. Consequently, the desirable degressive airbag pressure profile which has a force limiting effect cannot be implemented either with a two-stage gas generator alone or in conjunction with a large airbag vent. In conjunction with a large vent for force limitation, the mass flow must be significantly increased due to the major gas losses through the opened vent.

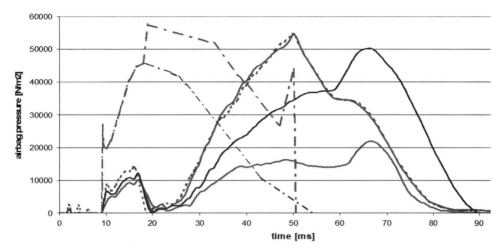

Figure 7.15: Mass flow profile for generating the degressive airbag pressure (blue: Airbag pressure of the standard system with standard mass flow (blue dashed); red: Airbag pressure with large vent and standard mass flow; violet dashed: Necessary mass flow in order to achieve a degressive pressure profile (violet) with a large vent;green dotted line: Airbag pressure in a degressive system with standard mass flow.

Compared to the standard system, 26 g (g force) instead of 46 g propellant will then be needed. The saw tooth characteristic resulting at the post-ignition of the second stage shows that at least three stages are required for the necessary mass flow profile. A very rapid limitation on the mass flow at the end (see the violet dashed curve in Figure 7.15) by suddenly stopping the gas supply (burn-out of the propellant or cold gas generator) is not technically feasible at this time.

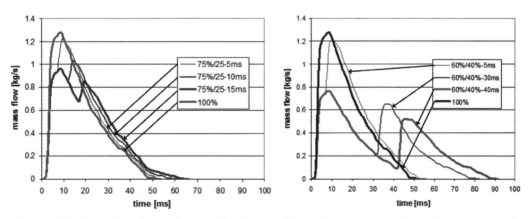

Figure 7.16: Examples of mass flow profiles in current two-stage gas generator systems compared to a one-stage system (100%); left: Distribution of the mass flow between the first stage (75%) and second stage (25%) with different time offset of the ignition of the second stage (in this example 5–15 ms); right: Distribution of the mass flow of the first and second stages at 60% and 40%, with a significantly longer time offset up to 40 ms.

7.3.2 Volume control

Control of the airbag volume tends to have a positive effect depending on the occupant's seat position; after all, it does influence the moment when the occupant interacts with the airbag (Figure 7.17). The functional profile of the restraint effect over the airbag pressure does, however, show a conflict of interest comparable to that of the mass flow (Figure 7.18). Accordingly, an early increase in the airbag pressure goes hand in hand with a significantly increased maximum airbag pressure, which may overload the occupant.

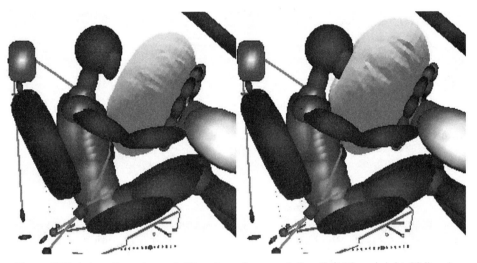

Figure 7.17: Interaction phase at 45 ms by volume variation (left 52 and right 68 liters).

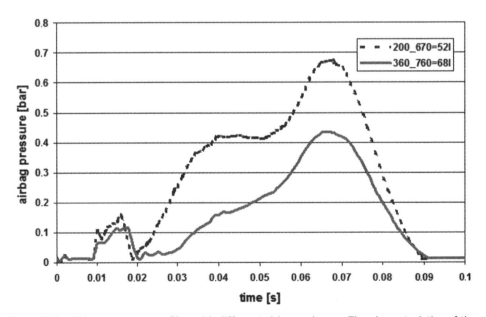

Figure 7.18: Airbag pressure profiles with different airbag volumes. The characteristics of the profiles are retained in spite of the variation in airbag volumes. The boundary curves of a small airbag with 52 liters (dotted line) and of a large airbag with 68 liters are shown.

A time-controlled volume increase, by providing additional volume, would also result in a pressure drop. Controlling the volume to achieve the desirable function profile in the restraint force consequently does not appear to be satisfactory. The airbag volume should be as large as possible, considering the occupant's seat position and out of position issues, to guarantee the earliest possible interaction of the bag with the occupant, and sufficient restraint travel for the airbag. This means that having an airbag size adjusted adaptively for the occupant in question would be helpful.

7.3.3 Vent control with constant pressure

Controlling the gas outlet openings or vents in the airbag is another possible way of controlling the airbag pressure. The quantity of gas emerging is controlled by a pressure-regulated valve. Pressure regulated valves are used in safety technology where they are referred to as safety valves. The switching times that can be implemented are approximately 2 ms. The vent diameter in this case must be have approximately a factor of three compared to a standard system in order to be able to generate a sufficiently rapid system response. If a particular pressure is exceeded, the valve opens, whereas it closes again if the pressure drops below that level. The plateau that is thus established in the airbag pressure profile has a force limiting effect (see Figure 7.19).

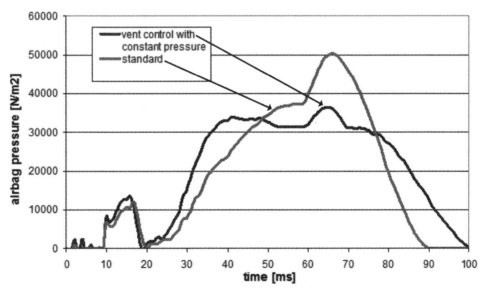

Figure 7.19: Principle airbag pressure profile of vent control with constant pressure of approximately 0.32 bar compared to the standard system. The brief increase in the pressure curve is due to the cessation of the steering column deformation. As a result of the delayed response characteristic, the safety valve is not able to compensate entirely for the resulting pressure rise.

Controlling the pressure according to the occupants and accident severity tends to have a positive effect. Interaction is improved because the vent is initially closed. The pressure limitation means that the force level to which the occupant is exposed is limited. However, the restraint force of the airbag is not constant when a constant airbag pressure level is selected. It increases as the contact surface with the occupant increases towards the airbag ($F_{Airbag} = p_{Airbag} \cdot A_{Occupant\ Airbag}$). This relationship means that a degressive airbag pressure profile is more favorable.

Vent control with a constant pressure level consequently results in a compromise. The pressure level must be high enough at the beginning of the crash in order to guarantee a good ride down effect, but subsequently low enough in the later crash phase to prevent transmitting the high vehicle decelerations to the occupant. This compromise leads to a failure to achieve an optimum pressure level and consequently to non-optimum occupant loading.

7.3.4 Airbag vent switching once

The airbag control has been implemented by means of two parameters, the vent diameter (e.g., outlet opening in the airbag has a choke point effect and is used for dissipating energy) and the vent opening time (e.g., the time at which the initially closed vent is opened with a specific vent diameter. The activation time of the vent, which is preferably opened pyrotechnically, is 2 ms). Varying both parameters for a load case (accident severity and occupant) leads to a minimum occupant loading (Figure 7.20 left). Figure 7.20 (right) shows the mode of effect of the two parameters. Toward the extremes of the selected range for the vent diameter (small to very large

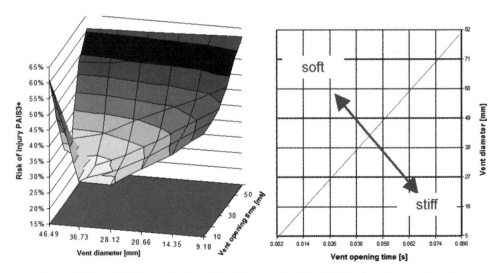

Figure 7.20: Left: The estimate of the injury probability PAIS3+ is influenced by the hardness of the airbag. The airbag hardness can be varied by the vent opening time and the vent diameter. Right: Principle relationship between airbag hardness and variation of the vent opening time/vent diameter.

vent), the airbag becomes too hard (high airbag pressure) or too soft (low airbag pressure). The behavior of the vent opening time can also be seen in comparison to the vent diameter. If the vent opens early, then the airbag is too soft. If the vent opens late, then the airbag is too hard. The minimum differs according to the load case, and consequently, induces an adaptively potential.

When the vent switches once with a comparatively large opening diameter, this means a degressive airbag pressure profile is created (Figure 7.21).

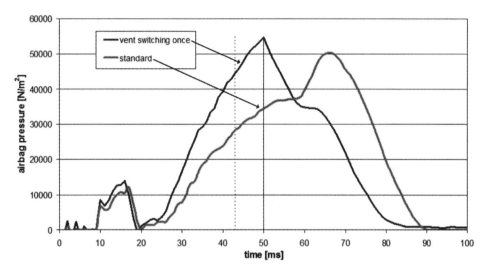

Figure 7.21: Airbag pressure profile of the degressive vent control (blue) compared to the standard (red). The plateau in the pressure drop in the adaptive system results due to the cessation in steering column deformation. In the standard, the steering column movement initially leads to a dip in the rise. Vertical lines: Switching time of belt force limiter (dashed) and switching time of vent opening.

This control is consequently referred to as degressive vent control. The degressive profile links an initially high level for improved interaction to a subsequently declining restraint effect, in order to be able to provide effective damping for the high deceleration forces which act as a result of the crash pulse. This means the restraint force acting on the occupant can be influenced favorably. This can be demonstrated by the plateau-shaped chest acceleration (Figure 7.22). However, if the pressure curve drops too strongly, this increases the risk of a puncture. A robust configuration can prevent this. A robust configuration can be achieved by including a safety margin travel, in which case the available forward displacement is not completely utilized, for example, by selecting a smaller vent diameter.

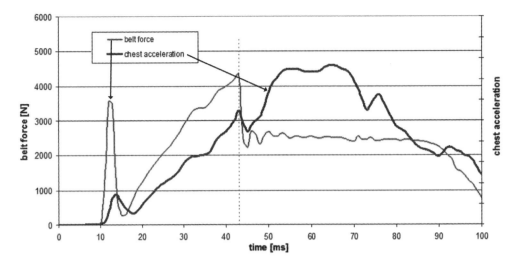

Figure 7.22: Chest acceleration profile (blue profile) of the degressive vent control. The degressive pressure drop in the airbag (see Figure 7.21) leads to plateau formation. Switching of the belt force limiter (dashed line) leads to a significant dip in the belt force (green profile) and a temporary dip in the chest acceleration (dashed line).

7.3.5 Switchable belt force limiter

The belt force is currently controlled by force limitation to a level of 4–6 KN. Some units referred to as degressive force limiters are installed in vehicles that require a higher belt force level briefly at the beginning by means of a break-loose moment until the force limiter takes effect and the lower level is set. In addition, some of the force limiters fitted in vehicles already have a switching function. This development has been promoted in order to be able to offer a certain level of adaptivity for different occupants, as well as to obtain travel-dependent switching of an improved degressive belt force limiter. The advantage of the degressive characteristic is, as in the case of the airbag system, that the initial interaction takes place rapidly at an elevated force level and the force level is reduced before the occupant loading reaches an excessive value. With a constant level, the two requirements for a steeper force rise and exploitation of the forward displacement cannot be achieved.

The necessity for belt force switching in conjunction with airbag vent opening is shown in Figure 7.23 (blue curve). In the case of a belt with a high force level (red curve), the available restraint distance is not used sufficiently. At a low belt force level (green curve), on the other hand, the initial interaction is insufficient, resulting in a significant rise in chest acceleration later.

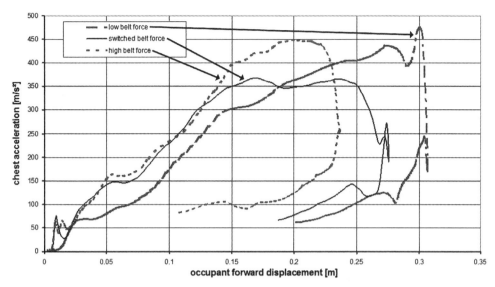

Figure 7.23: Comparison between the chest acceleration/travel characteristics acting on the occupant at a high belt force (red dotted), low belt force (green dashed) and in switching the belt force from the high level to a low level (blue).

7.3.6   Comparison between the airbag control concepts

A selection of possible control concepts for influencing the restraint function acting on the occupant have shown that the airbag vent control and belt force control can vary the functional profile of the restraint effect most effectively to guarantee a trapezoidal characteristic for occupant deceleration. The significant differences in potential are due to the different restraint effects of the system controller. In contrast to the other airbag control concepts, vent control can offer a degressive force function. This degressive behavior initially leads to improved interaction. The force limitation is achieved by the rapid pressure drop after vent opening. This degressive force behavior is also sensible with regard to the belt force, and can be implemented adaptively using a switching belt force limiter.

7.4   Occupant and accident severity-specific adaptivity

Adaptivity can offer an optimized system for each individual load case. The existing conflict of interest involving configuring one system which is the same for several conditions might consequently be resolved in principle.

7.4.1   Airbag vent control with switching belt force limiter

In principle, the airbag control concept of degressive vent control with switching belt force limiter allows the required adaptive adjustment in restraint effect to be achieved by providing different airbag hardness according to the occupants and accident severities.

The potential of an adaptive restraint system with variability levels can be demonstrated by simulation for different occupants and accident severities (crash pulses). The individual loadings (head acceleration and chest acceleration) are shown in Figure 7.24, indicating that, with an adaptive airbag vent control, it is possible to reduce the total loading not only for one occupant type, but for a variety of occupants. These are described chiefly as the 5-percentile, 50-percentile, and 95-percentile occupants. As shown in Figure 7.24 individual restraint can reduce the differences in load between several occupants. Moreover, an airbag vent control can provide a further decrease of overall load on the 50 percentile dummy. This effect results from a more rectangular characteristic of the restraint function.

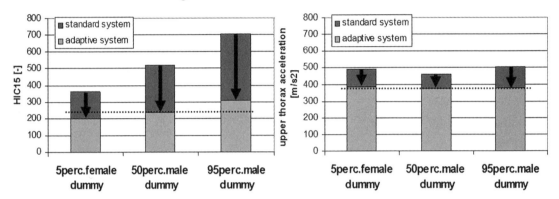

Figure 7.24: Comparison between the standard restraint system and the adaptive restraint system in the US-NCAP crash test. There is a homogenization of the different loading levels on the occupants (horizontal line).

In order to provide robust triggering of adaptive systems, it is necessary to have suitable sensors that are able to classify not only the accident severity, but also the occupants sufficiently well. The accident severity can be determined by using the signal for vehicle deceleration from the airbag control unit. To do this, the signal for vehicle deceleration is also used after the restraint systems are triggered until the vent opens. Classifying the occupants means it is necessary to determine their positions so that the available forward displacement can be calculated in terms of energy. Also, their weights must be measured to take into account the required force level of the restraint system.

7.4.2 Self-adaptation of the belt force for different occupants

One possible joint setting of the belt function is similar to the steering column deformation setting as a result of the self-adaptation of the belt force acting on the occupant. The extraction speeds of the belt strap out of the belt retractor vary according to the different weights of the occupants. As a result, there is a different wrapping friction between the belt deflector and belt webbing, which means the belt force acting on the occupant is different as well. Figure 7.25 shows the relationship for the standard system and the adaptive vent control as an example. In a switching belt force limiter, the 5% dummy has a break-loose moment which is initially up to 1.5 KN lower than the 50% dummy. This leads to a belt force level which is about 0.5 KN lower due to the different wrapping friction. In the standard belt system, the belt force level on the occupant is about 1.0 KN lower throughout the entire profile.

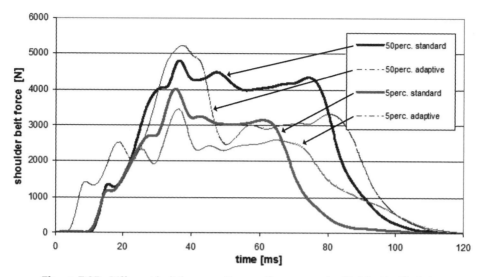

Figure 7.25: Different belt forces acting on the occupant with identical belt force limitation of the automatic belt retractor.

7.4.3 Potential for the occupant without seat belt

The possible potential for an occupant without a seat belt was analyzed for the load case FMVSS208-30mph using two-stage adaptivity (constant airbag geometry and variable vent opening time and vent diameter). The adaptive airbag control can also be significantly effective even for an occupant without a seat belt [7-5]. If the occupant is not wearing a seat belt, the initially closed vent of the airbag control means that a particularly high level of energy is converted into a high airbag pressure. This holds true with a low occupant weight and low accident severity. Here too, it is shown that the injury risk can be significantly influenced by the thorax loading. The magnitude of the airbag pressure is sufficient to restrain the relatively low mass of the head, whereas it is not sufficient for the mass of the thorax.

However, failure of the occupants to wear seat belts also results in a very high loading on the airbag system. The airbag pressure increases markedly. The dominant influence of the occupant's weight becomes particularly clear, because in this load case, no energies can be converted by the belt system. The self-adaptation of the belt system forces because of different occupants and seat positions (belt absorbs a greater proportion of the energy with the seat in the rear position) is no longer provided. This means the airbag system is overloaded, which correlates with accident investigations [7-10] that demonstrate that the belt is the most effective means of restraint. As a result, failure to use the belt in the standard system results in high occupant loadings, even with an accident severity of about 18.6 mph (30 km/h). Adaptive airbag control can partially compensate for the increased risk of injury if the seat belt is not used. From an accident severity Δv of approximately 31 mph (50 km/h) onwards, however, even an adaptive system can no longer prevent an increased risk of injury. To this extent, use of a belt is always advisable even if an adaptive restraint system is installed in a vehicle.

7.4.4 System function and potential for the front-seat passenger

The front-seat passenger is involved in accidents at a statistical frequency of about 30%. It is apparent that the concept of vent control and its function for optimizing restraint are also suitable for the front-seat passenger.

The five-stage adaptivity with constant airbag geometry allowed its potential for the front-seat passenger to be estimated. Figure 7.26 shows the reduction in PAIS3+ injury probability and the reduction in head and thorax injury probability on average over six load cases (three occupants: 5-percentile, 50-percentile and 95-percentile evaluated under two accident severities: USNCAP, EuroNCAP). Without adapting the front-seat passenger airbag geometry, adaptive vent control itself can significantly improve the front passenger injury risk by about 30%. Adapting the front-seat passenger airbag geometry to the adaptive vent control implemented with the driver could result in a further increased potential for injury reduction.

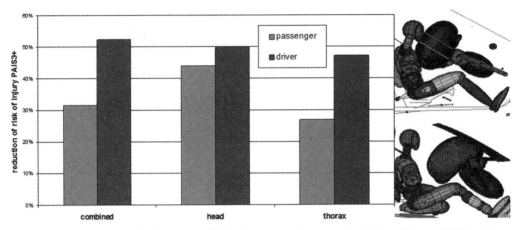

Figure 7.26: Reduction in the injury risk for the thorax, the head and the estimate PAIS3+ of the front-seat passenger compared to the driver. Right figure: Different interaction between the driver (top) and front-seat passenger (bottom) 40ms after the beginning of the crash.

The different restraint effect between the driver and front-seat passenger is apparent. At 40ms after the beginning of the crash, the driver (in this example, a 50% male driver) is already interacting with the airbag, while the front-seat passenger is still positioned relatively far away from the deploying airbag. Interaction of the airbag with the front-seat passenger's thorax occurs later than that of the driver, because the large front-seat passenger airbag cannot be deployed so close to the seated occupant because of out of position issues.

As far as the front-seat passenger is concerned, the protective potential of the airbag is directed primarily at the head because the available airbag restraint travel is larger because of the higher volume of the airbag (e.g., there is no steering wheel to prevent the forward movement of the front seat passenger). The potential of airbag control is also revealed in the case of the thorax, because a reduction in the belt force at the same time as an increase in the active surface area over which the airbag force acts (concept of *less belt–more airbag*) can be implemented over a relatively large airbag restraint travel. In the load case of the occupant without a seat belt, the potential is revealed even more prominently because of the airbag size compared to that of the driver.

7.4.5 Summary of adaptivity

The adaptivity of restraint systems may offer an optimized system for individual load cases. The existing conflict between configuring one system to cover several conditions can thus be dispensed with in principle. The greatest adaptivity potential can be achieved by considering the total of seven relevant input parameters for the restraint system: airbag vent switching, airbag opening time, width of volume control, diameter of volume control, two different belt load limiter levels, and switching time. As the variability of the adaptive restraint system is reduced, so are both the complexity and the effectiveness of the system. The necessary simplification of the system complexity showed that airbag volume control does not have a significant influence.

Further system simplification through common setting of the belt function has already shown a significant reduction in effectiveness. In the implementation of the technical system, the reduction in effectiveness does appear acceptable, however. Ultimately, it was possible to show that a vent that switches once and has a constant vent cross section can achieve a remaining effectiveness of about 34% compared to the standard system [7-6]. It is also shown, however, that no adaptivity, meaning only optimizing the degressive vent control in one load case over the desired trapezoidal characteristic does not provide an acceptable solution, because, in some load cases, there can even be a significant impairment compared to the standard. To this extent, optimization of the restraint effect by degressivity can only be implemented sensibly by adaptivity.

7.5 Estimate of the potential of adaptive restraint systems in a real accident

Calculating the benefit of adaptive restraint systems in a real accident situation is difficult, because converting simulated loading parameters for the occupants to real human beings is by no means straightforward [7-11], [7-12], [7-13]. This chapter develops an approach which moves away from the problematical evaluation of individual injuries. This means that the potential of an adaptive restraint system determined by simulation techniques can be transformed to the real accident situation.

7.5.1 Injury probability

Based on the process of the US-NCAP, described in [7-14], the protection criteria for potential serious head and thorax injuries (AIS3+) are compared to one another. The relationship between the protection criterion and injury probability has been established in biomechanical investigations and reflects the averaged human loading capacity.

However, this loading is perceived differently by each human being. If the loading is greater than the individual loading capacity, then an injury will result. However, evaluating the individual loading capacity does present problems, because this can vary depending on the age of the occupant and perhaps on other individual characteristics that cannot be established, such as the physical condition, muscle mass, and muscle tension. As a result, very wide deviations can occur in individual cases. The approach involving determination of injury probability is highly complex and can only be represented as a simplification depending on the acting loading; however, it does offer a significant level of statistical significance under the assumption that the influence of the individual properties tends to be cancelled out by the injury probability on average.

The accuracy of determining injury probability is initially in the forefront of the optimization strategy, because it does not influence optimizing the individual injury parameters. The accuracy is only important for a desirable estimate of the potential in a real accident situation. For the head, the probability is calculated on the basis of the HIC according to [7-14]:

$$P_{HeadAIS3+} = \frac{1}{1+e^{5,02-0,00351 \cdot HIC}}.$$ (7.4)

The upper limit value of the HIC at 1000 corresponds to an AIS3+ injury probability to the head of 18% in accordance with the formula below (see Figure 7.27). The steepest point on the curve lies between the HIC values 1000 and 2000. There is a 90% probability of injury at an HIC of 2050, which correlates to exploiting about 200% of the limit value.

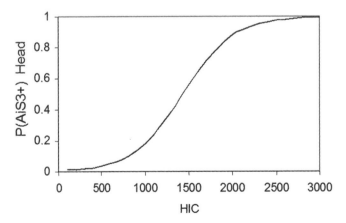

Figure 7.27: Probability of AIS3+ injuries to the head.

For the thorax, the probability is calculated on the basis of the combined thorax injury (CTI), a single parameter that combines the chest acceleration (3ms) and the chest compression according to [7-15]:

$$P_{ThoraxAIS3+} = \frac{1}{1+e^{7,529-6,431 \cdot CTI}}.$$

(7.5)

This is shown in Figure 7.28. In this case, the probability of 25% corresponds to the limit value of 1. At 150% of the limit value, the probability is 90%. The rise in the injury probability for the CTI criterion is consequently steeper than for HIC.

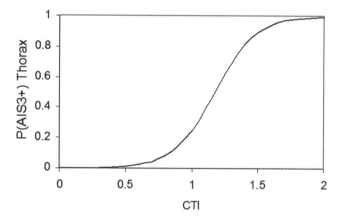

Figure 7.28: Probability of AIS3+ injuries on the thorax.

The individual probabilities for the head and thorax do not make it possible to derive a precise valid total probability on the basis of the formula below, due to the lack of independence. The relationship according to Formula 7.5, however, does provide a very effective possibility for a balanced optimization of the significant parameters (see Figure 7.29):

$$P_{total} = 1 - (1 - P_{Head}) \cdot (1 - P_{Thorax}).$$

(7.6)

This optimization and evaluation strategy has proven to be effective in technical simulation investigations. These have shown that machine optimization can be achieved without prejudicing other loading parameters that are not considered in the total injury probability. Further advantages of the evaluation concept are concerned with the higher weighting of the chest loading and the ability to use automatic calculations. In order to estimate the potential in a real accident situation, it is not possible to use Formula 7.5 directly because of the lack of independence of the injury probabilities, consequently the estimator *PAIS3+* has been introduced (Formula 7.6). The imaginary estimator is calculated as follows by taking all serious MAIS3+ injuries into account:

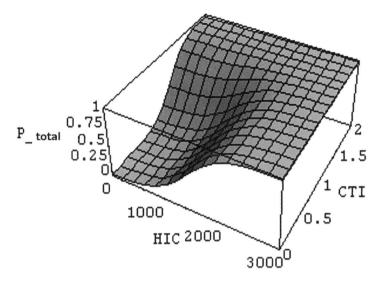

Figure 7.29: Balanced optimization of the significant parameters by forming the total injury probability.

$$PAIS3+ = 1 - (1 - P_{ThoraxAIS3+}) \cdot (1 - P_{HeadAIS3+}). \tag{7.7}$$

7.5.2 Principal problems with transferring the potential determined in simulation techniques to real accident situations

The accident data of the VW-GIDAS database makes it possible to calculate not only the probability of injury to individual bodily regions, but also the total injury probability by statistical means.

In the simulation, individual occupant loadings can be determined by parameters such as the HIC and converted into injury probabilities for bodily regions by means of biomechanical relationships. However, this does not make it possible to calculate injury probabilities, because there is no statistical independence between individual injury probabilities. Given statistical independence of the individual injuries, the following (Formula 7.8) would apply on the basis of the individual probabilities p_i for P_{total}:

$$P_{total} = 1 - \prod_{i=1}^{n} (1 - p_i). \tag{7.8}$$

Given the statistical dependency of the individual injuries on the probabilities p_i in the accident data, P_{total} can only be estimated within the possible limits [7-16]:

$$\max_{i=1,2,..n} [p_i] \le P_{total} [p_1, p_2 \dots p_n] \le \sum_{i=1}^{n} (p_i) \tag{7.9}$$

Consequently, calculating P_{total} can only be done as an estimate within a wide bandwidth, and does not lead to an acceptable solution. However, even given knowledge about the total injury probability from the simulation, converting this into the accident situation would be problematic because the transformation function is not known. Therefore, it is necessary to use a method that is not dependent on the problematic conversion on the basis of injury criteria and is also not dependent on the total injury probability that has to be determined by the simulation.

One possible measure is to convert the potential determined in the simulation into a reduced accident severity. In this case, the reduced occupant loadings are not calculated as described above by the injury parameters and injury probabilities, but are first converted into the equivalent accident severity.

7.5.3 Equivalent accident severity

The reduction in loadings on the occupant by using adaptive restraint systems can be compared to a reduction in accident severity, given the current systems. This approach offers the possibility of converting the effect obtained in the simulation into the accident situation. The VW-GIDAS accident database was used for describing the representative accident situation in Germany [7-17], [7-18].

Adaptive occupant protection is intended to improve thorax protection. Consequently, it appears sensible at first to use the thorax injury probability parameter. To do this, the loadings on the three representative occupant types (5%, 50%, and 95% dummy) were determined by simulation for the adaptive system and the standard system in the case of different crash pulses. The main emphasis was placed on technical feasibility, which is why the single-stage adaptive system was selected. The result is shown in Figure 7.30 as a diagram of the injury probability

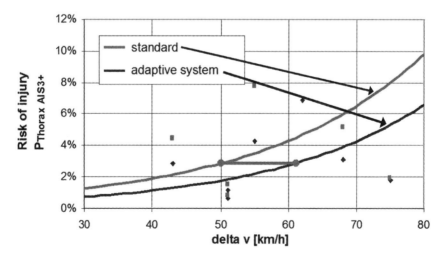

Figure 7.30: Determining the reduced thorax injury probability of the adaptive system compared to the standard system in the equivalent accident severity (Δv) (horizontal bar); The reduction in loading can be converted in its effect as a reduction in accident severity (Δv).

against accident severity Δv. Exponential function profiles were selected because these correspond to the characteristics of the real accident data. The significant distribution of individual calculation points comes about because of the deliberate choice to select different magnitudes of vehicle decelerations with the same Δv. The influence has already been discussed.

Determining the equivalent accident severity (Δv) of the adaptive system compared to the standard system is a result of considering the same thorax injury probabilities. In the case shown in Figure 7.31 (horizontal bar), the risk in the standard system with an accident severity of approximately 31 mph (50 km/h) corresponds to a comparable risk in the adaptive system with an accident severity of as much as approximately 37.9 mph (61 km/h).

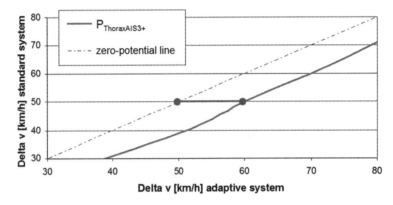

Figure 7.31: Conversion of the speed-equivalent injury probability P$_{ThoraxAIS3+}$ of the adaptive system against the standard system. The diagonal dashed line is the zero potential line at which the adaptive system would have the same accident severity as the standard system. The reduced thorax injury probability of the adaptive system compared to the standard system in the equivalent accident severity (Δv) is shown as a horizontal bar for a sample point.

In order to simplify the representation, the reduced thorax injury probability from the adaptive system compared to the standard system (Figure 7.30) is transformed below into the equivalent accident severity (Figure 7.31). This applies to all three identified dummy types, because a comparable effectiveness could be established by varying the crash pulses.

Furthermore, the equivalent accident severity can be calculated for additional parameters. There are different reductions in the accident severity for the loading parameters HIC, CTI, and parameters derived from these, such as the injury probability to the head and thorax. The tendency of the curves is identical in all parameters. The potential can be estimated from a conservative approach taking the minimum reduction that can be observed in all considered parameters. The potential increases slightly with greater accident severity. On average, an adaptive system can reduce the effective accident severity by up to 6 mph (10 km/h). In the diagram (Figure 7.31), this is shown by the distance from the dashed diagonal line (zero potential line).

7.5.4 Calculation of the virtual injury distribution in the field

The conversion of the existing injury distribution in real accident situations into a virtual injury distribution by using an adaptive system is performed using the reduction in accident severity obtained by simulation techniques (see previous chapter). Figure 7.32 shows the distribution of the probability of suffering serious MAIS3+ injuries for the accident situation and the virtual distribution using an adaptive restraint system.

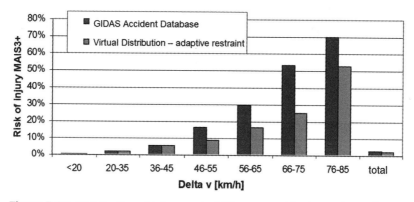

Figure 7.32: Distribution of injury probability over accident severity Δv for the accident situation and the virtual distribution by using an adaptive restraint system. The lower accident severity classes Δv up to 45 km/h were not considered in the simulation. Although it can be assumed that there is also potential for improvement here as well.

The virtual distribution results from the reduction in accident severity determined by the simulation. In the accident severity classes, the new virtual accident severity was determined for each accident and then the new distribution was made. The highest effectiveness can be achieved in the accident severity classes 34.8–40 mph (56–65 km/h) and 41–46.6 mph (66–75 km/h). Due to the high proportion of collisions with low accident severity (classes < 12 mph [20 km/h] and 12–18.6 mph [20–30 km/h]), there is, on average, a low overall injury probability.

7.5.5 Effectiveness and benefit of adaptive restraint systems in the field

Calculating the total effectiveness on the basis of probabilities is not common in probability calculation, but in the current case it does represent a sensible parameter and is frequently used in accident research. From Figure 7.32 it is possible to determine the effectiveness of adaptive restraint systems in a real accident. The calculation can be performed both using the individual accident severity classes and the total for the accident situation. The total effectiveness for reducing the risk of MAIS3+ injury is calculated as follows:

$$Eff_{totalMAIS3+} = \frac{P_{FieldMAIS3+} - P_{VirtualMAIS3+}}{P_{FieldMAIS3+}} \qquad (7.10)$$

The effectiveness for MAIS3+ is calculated, using Formula 7.10, as 25.7%. This means that having an adaptive restraint system could avoid approximately 25% of seriously injured car occupants in frontal collisions. The effectiveness's for additional injury severities can be calculated by using the procedure described above. The effectiveness for injury severities MAIS3+ to MAIS5+, based on the conservative estimate using the VW-GIDAS accident database, can be seen in Table 7.1.

Table 7.1: Effectiveness by using an adaptive restraint system for different injury severity classes in frontal collisions			
Injury severity	MAIS 3+	MAIS 4+	MAIS 5+
Effectiveness of adaptive restraint systems	25.7%	24.6%	15.9%

Taking the simplified assumption that injury class MAIS5+ principally includes fatalities (the MAIS5+ injury class includes MAIS5 and MAIS6 injuries; the lethality rate of these injuries is very high), means a reduction of about 16% in fatalities and a reduction of about 25% in serious and very serious injuries in frontal collisions when an adaptive restraint system is used.

7.6   References

7-1   Gonter, M., Zobel R., Spies, U. Potential of adaptive restraints in frontal collisions. 7th International Airbag Symposium on Car Occupant Safety Systems, 2004.

7-2   Schramm, C., Fürst F., van den Hove, Gonter, M. Adaptive Restraint Systems— The Restraint Systems of the Future. 8th International Symposium Airbag 2006, December 4–6, 2006, Karlsruhe, Germany.

7-3   Ragland, C.L. Evaluation of crash types associated with test protocols. ESV Paper No. 339, 2003.

7-4   Stucki, S.L. Hollowell, W.T., Fessahaie, O. Determination of frontal offset test conditions based on crash data. NHTSA Paper Number 98-S1-O-02, 1998.

7-5   Gonter, M. Steuerkonzepte für adaptive Airbagsysteme zur insassen- und unfallspezifischen Optimierung des Insassenschutzes. ProBusiness Verlag, Berlin, 2007.

7-6   De Leonardis, D.M., Ferguson, S.A., Pantula, J.F. Survey of driver seating positions in relation to the steering wheel. SAE Paper 980642, 1998.

7-7   Austin, R.A., Faugin, B.M. Effect of vehicle and crash factors on older occupant injuries. ESV Paper No. 102, 2003.

7-8   Zhou, Q., Rouhana, S.W., Melvin, J.W. Age Effects on Thoracic Injury Tolerance. SAE Paper Number 962421, 1996.

7-9   Foret-Bruno, J.Y., Trosseille, X., Le Coz, J.Y. et al., Thoracic Injury Risk in Frontal
      Car Crashes with Occupant Restrained with Belt Load Limiter. SAE Paper
      Number 983166, 1998.

7-10 Zobel, R. Benefit from fleet change and restraint systems. 9. EMN Congress
      Bucharest, Romania, May 20–22, 2004.

7-11 Cuerden, R., Hill, J. Kirk, A., Mackay, M. The Potential Effectiveness of
      Adaptive Restraints, IRCOBI Conference, 2001.

7-12 Kent, R., Matsuoka, F. et al. Development of an age-dependent thoracic injury
      criterion for frontal impact restraint loading. ESV Paper No. 72, 2003.

7-13 Yajima, H., Nozaki, A., Influence on Injury Criteria of Occupant Restraint
      System. SAE Technical Paper Series 2000-01-0623, 2000.

7-14 Kramer, F. Schutzkriterien für Fahrzeug-Insassen bei Frontalkollisionen,
      Verkehrsunfall und Fahrzeugtechnik. Heft 7/8, 1992.

7-15 Kleinberger, M., Sun, E., Eppinger, R. et al. Development of improved injury
      criteria for the assessment of advanced automotive restraint systems. NHTSA,
      www.nhtsa.dot.gov, 1998.

7-16 Angewandte Statistik, Springer Verlag, 10. Auflage, Ungleichung nach
      Bonferroni, Seite 39, 2002.

7-17 Becker, H., Donner, E., Graab, B., Zobel, R. In-Depth-Erhebung und
      Einzelfallanalyse. Werkzeuge zur Verbesserung der Fahrsicherheit
      im Volkswagen-Konzern. 1. Dresdener Tagung Verkehrssicherheit
      Interdisziplinär, 2003.

7-18 Brunner, H., Georgi, A., Sachs, L. Drei Jahre Verkehrsunfallforschung an der
      TU-Dresden. Automobiltechnische Zeitschrift (ATZ). 2003.

Chapter 8
Compatibility of Passenger Cars, Trucks, and Pedestrians

8.1 General

Because safety measures like seat belts have reduced injuries in single car accidents further improvement in traffic safety is focused on compatibility between all traffic participants. The first research work in this area dates from the early 1970s [8-1, 8-2, 8-3].

To understand the total accident scene the following traffic participants have to be considered: pedestrians, bicycles and motorcycles, passenger cars, trucks, and buses. Injuries can occur in single car accidents or in collisions with other traffic participants and other obstacles.

Compatibility investigations have shown that a global approach that includes all traffic participants is very complex and solutions are not easy. If compatibility issues are included in the design of vehicles the following parameters are important in the collision sequence:

- Mass ratio, geometry, and force-deformation characteristics of the vehicle structure and installation and mass of the power train unit
- Restraint system and geometry of the occupant compartment
- Performance of interior parts like the steering system, dashboard, and upholstery.

Compatibility analyses have to take all relevant factors into consideration including the reduction of the accident costs versus potentially reduced comfort and increased energy and raw material consumption, although today's vehicle owners don't want to sacrifice the protection of their own vehicle as a consequence of compatibility measures. The description of the accident type and a correlation to the type and severity of injuries is needed for the optimization process as well as a correlation between the injury severity (AIS-rating) and the criteria, which are measured and calculated with the help of dummies. Only then can measures in the vehicles and road infrastructure be evaluated.

In different vehicle masses and restraint systems the force and deflection characteristics could be changed in such a way as to minimize the total accident costs. The following information is necessary:

- The accident in correlation to the AIS-injury severity
- The correlation of the AIS to the measured data of the test dummy
- The financial assessment of the occupant injury
- The higher production costs created.

With this information the vehicle safety measures could be optimized in the direction of minimum total costs [8-4]. In the last few years scientists and vehicle manufacturers, as well as insurance companies, have investigated the problem of compatibility in greater depth [8-5]. Newer investigations confirm the necessity to use mathematical models [8-6] to judge more precisely the compatibility performance of vehicles, for example, to avoid local deformation force peaks, which helps increase the compatibility in car to car crashes.

Therefore, FEM models and/or deformation force measurements during a frontal collision can define a force distribution quality as a scale for compatibility. In connection with this, the bulk head concept is a very interesting solution [8-7]. The basic assumption for this concept assumes, that if the occupant compartment is rigid enough, the performance of the today's restraint systems will allow an occupant to survive even in severe vehicle to vehicle crashes.

After more than 30 years of research and development work, the requirements are becoming more concrete. The 3306.9 pound (1500 kg) barrier, defined through the IIHS with a higher deformation element, is driven at 31 mph (50 km/h) under an angle of 90° into the side of the tested vehicle. The front to front compatibility requirements on the geometric design are defined. The height of the deformation force and a height evaluation are also discussed [8-8, 8-9]. These requirements also cover light duty trucks and SUVs. First positive results are already reflected in accident statistics. The modification of the FMVSS 214 promises further improvements, for example, through the limitation of the Aggressivity Average Height of Force (AHOF)

8.2 Passenger car/truck collisions

Collisions between trucks and passenger cars present a specific compatibility problem because the mass ratios and especially the geometric differences are critical. This is true not only in Europe but also in the U. S. where the compatibility between light duty trucks and passenger cars is a critical issue.

For the truck, a future solution called the plank frame as shown in [8-10], provides underside protection for the truck and trailer. As is often the case, such solutions can only be brought into the market if legal and/or insurance requirements support them.

8.3   Pedestrian collisions

A further incompatibility exists during a collision between pedestrians and passenger cars. The different mass ratios and body sizes and the physical resistance of the pedestrians provide little opportunity to improve the situation. In general the following measures could be positive:

- Reduced waiting time on light signal < 40 s
- Geometric design of pedestrian cross walks, which automatically position pedestrians to see the oncoming traffic
- Separated signal between car turns and pedestrians
- Increased awareness on the part of the driver of the influence of speed on braking distance
- Education, especially in kindergarten and elementary school, and teaching pedestrians to wear visible clothing
- Increased attention for special traffic situations

In many countries pedestrian protection has a high priority. In Europe, accident avoidance elements were integrated into the legislation related to pedestrian protection (EG 78/2009) [8-11]. Table 8.1 summarizes some of this legislation

The importance of pedestrian protection was increased because pedestrian safety is an integral part of the NCAP program. In addition to national legislation, the Global Technical Regulation seeks to achieve a standardization of the national legislation as shown in Figure 8.1.

Table 8.1: Legislation pedestrian protection publication VO(EG) No. 78/2009, February 2009 introduction dates			
NT = new models NR = ...	Brake assistant	Phase 1	New Phase 2
M1 ≤ 2.500 kg zGG	NT November 24, 2009 NR February 24, 2011	NT (since October 1, 2005*) NR December 31, 2012	NT February 24, 2013 NR February 24, 2018
N1(M1)** ≤ 2.500 kg zGG	NT November 24, 2009 NR February 24, 2011	NT (since October 1, 2005*) NR December 31, 2012	NT February 24, 2013 NR February 24, 2018
all M1	NT November 24, 2009 NR February 24, 2011		NT February 24, 2015 NR February 24, 2019
all N1	NT February 24, 2015 NR August 24, 2015		NT February 24, 2015 NR February 24, 2019

Laws: Global Technical Regulation (GTR)

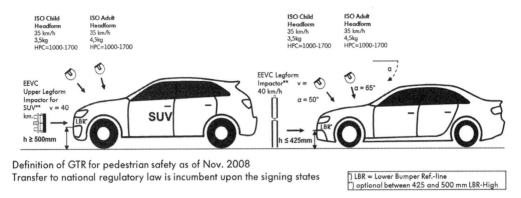

Definition of GTR for pedestrian safety as of Nov. 2008
Transfer to national regulatory law is incumbent upon the signing states

*) LBR = Lower Bumper Ref.-line
**) optional between 425 and 500 mm LBR-High

Figure 8.1: Global Technical Regulation to achieve a standardization of the national legislation.

A number of front end components like bumpers, spoilers, coolers, headlights, and lock and hinge fenders, as well as the A-pillar and windshield, wiper shafts, and components of the powertrain directly below the front hood could be adapted to provide pedestrian protection from a colliding vehicle. An appropriate sensor system would also be needed to actively move the front hood upward to get enough free deformation space to absorb the energy of the pedestrian's head.

8.4   References

8-1   Ventre, Ph. Homogenous Safety and Heterogeneous Car Population, 3. International Technical Conference on Experimental Safety Vehicles. Washington, D.C. June 1972.

8-2   Seiffert, U. Probleme der Fahrzeugsicherheit, Dissertation. Technische Universität Berlin. 1974.

8-3   Appel, H. Sind kleine Wagen unsicherer als große?. VDI-Nachrichten. Düsseldorf, Nr. 7/14.02. 1975.

8-4   Richter, B., et al. Entwicklung von PKWs im Hinblick auf eine volkswirtschaftlich optimalen Insassenschutz. Abschlussbericht BMTF, S. 116, gefördert vom Bundesminister für Forschung und Technologie. 1984.

8-5   Huber, G. Passive Safety of Vehicles including Partner Protection. FISITA. Prague. 1996.

8-6   Relou, J. Dissertation Methoden zur Entwicklung crash-kompatibler Fahrzeuge. TU Braunschweig. 2000.

8-7   Schwarz, T. Selbst- und Partnerschutz bei frontalen PKW-PKW-Kollisionen (Kompatibilität), VDI-Bericht. Reihe 12, Nr. 502, VDI-Verlag. Düsseldorf, ISBN-3-18350212-7.

8-8   Scheef, J. Anforderungen an Fahrzeugsicherheit. Symposium Concept, Graz. Februar 2004.

8-9   Insurance Institute for Highway Safety, Status Report Side Impact Crash Tests. Vol. 39, No. 5, April 2004.

8-10 Schimmelpfenning, K.H. Bord Frame, a Possible Contribution to Improve Passive Safety. 15. International Technical Conference on the Enhanced Safety of Vehicles. Melbourne, Australia. May 1996.

8-11 http://wg17.eerc.org/wg17publicdoes/wg17publication.html.

Chapter 9
Calculation and Simulation

9.1   Introduction

Toward the end of the 1960s, governmental requirements pertaining to motor vehicle and road traffic safety increased dramatically. At the same time, computer systems were too slow to respond in a timely way to the questions from development engineers about the behavior of structures. Ideally, answers to questions about the effects of design changes would have been provided within 24 hours. Supercomputers were not able to help to solve design questions until the beginning of the 1980s. One of the pioneers in this area was Mr. Seymour Cray from the U. S. His computers were also used in Europe and had the requisite computer power to provide answers to questions posed by the development engineers within an acceptable period of time.

But it was not only in the field of crash simulation that substantial progress was made. Computer simulations proved very beneficial in matters related to numerous aspects of accident avoidance during the vehicle development process. Of course, substantial advances were made in many other areas with the application of new methods for calculation and simulation, including vehicle handling, comfort, aerodynamics, powertrains, electronics, ergonomics, emissions, and crash management as well as accident reconstruction and analysis.

9.2   Man-machine-interface

Effective development and design processes require the integration of the human-machine-interface to accommodate the needs of the driver in a very early stage of the development-process [9-1]. In addition to tests in real driving situations, the use of simulation tools is gaining increasing importance.

Supplementing subjective seat of the pants assessments, simulations provide valuable opportunities for repeatable objective evaluations under a variety of conditions and circumstances. Based on cognitive science, driver ergonomics related to learning and interactional behavior provide new approaches to optimize and create increasing transparency to control logic and system design.

In the meantime, numerous driving simulators are used to evaluate and optimize the potential of different control and display components. To get a most realistic reaction from the driver fixed seat and dynamic simulators are used.

The dynamic driving simulator can be either a separate test device, as shown in Figure 9.1 or a modified vehicle [9-2] in which various parameters can be changed during driving operations, and in some cases, the vehicle can be remotely operated from a special cabin. With this special car it is possible to test and evaluate different assistance systems, to a limited extent in real traffic or without such restrictions, in special proving grounds. This integrated process allows questions regarding the objective benefit of these concepts to be assessed together with the acceptance by the potential customers, Figure 9.2 [9-3].

Figure 9.1: DLR-Driving Simulator [9-1].

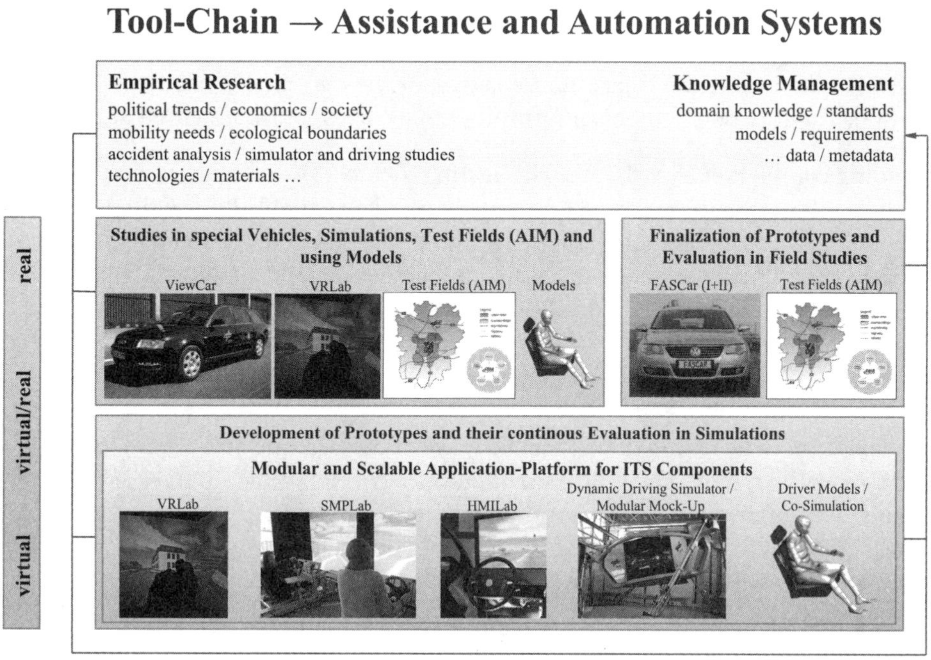

Tool-Chain → Assistance and Automation Systems

Figure 9.2: Integrated system design [9-3].

9.3 Computer-aided development process by HIL, VIL, SIL

9.3.1 Simulation of predictive safety systems

It is not enough to assure accuracy in the implementation of new concepts and systems, it is also necessary to verify feasibility and operational reliability of a safety function under real-world conditions. To do this, especially with regard to safety features and functions, detection and quantification of false alarm rates on the basis of reliable, representative data and extrapolation is a prerequisite. Due to the quantity and breadth of the large number of measurement data required, and the complexity of the verification procedure, a computer-aided development process is not just nice to have, but rather is an absolute necessity. Real measurement data are recorded and evaluated in multiple iteration loops, while at the same time, algorithms are developed for application of the function itself. In most cases, development of the function takes place either as model-in-the-loop (MIL) or software-in-the-loop (SIL). In addition, logical calculation rules necessary for the function are derived based on synthetic or simulated data. The advantage in this development step is that the logical component is essentially decoupled from the time component, enabling large quantities of data to be evaluated in the shortest possible time. Several thousand simulated collision scenarios have been used for algorithm development and optimization in the case of the pre-crash functions referenced above.

In subsequent development steps, the influence of real, connected control units and real (real-time) behavior is established in the hardware-in-the-loop (HIL) test.

Components such as bus communication and prototype control units, as well as actuator system behavior, are analyzed and sub-functions are checked. It must be possible to determine the time jitter behavior for the real-time capability of a system function. The function is then configured with a corresponding degree of tolerance.

Following completion of MIL/SIL/HIL testing, flexible system software that has already reached a sufficient level of maturity will be available for the final application step dealing with complete function in the vehicle itself. By means of various parameter sets, it is possible to apply a conservative, neutral, or pro-active system and function behavior directly in the overall vehicle environment in real-world driving scenarios.

9.3.2 Vehicle-in-the-loop

In the past, there were only two possible ways of researching and testing new functions in the rapidly growing area of automotive electronics. Driver assistance systems that support the driver in potentially critical traffic situations were tested in experiments involving foam cubes and other objects, such as specially protected target vehicles, or in driving simulators.

On one hand, traffic situations in which the assistance systems support the driver are becoming more complex, and it is becoming increasingly difficult to find reproducible setups for real vehicles. On the other hand, the haptic and kinesthetic concepts currently in the pipeline for human/machine interfaces make it almost impossible to use stationary simulators, or, alternatively they demand the same quantity of realism as moving simulators. The most modern driving simulators are still imperfect and lead to results that can only be transferred into reality with substantial difficulty, including problems such as motion sickness suffered by some test subjects.

Vehicle-in-the-loop (VIL) testing and the simulation environment developed by Audi AG [9-4] opens up an enormous breadth of new possibilities by combining the best of both worlds, i.e., the safety, reproducibility, and flexibility of driving simulators, with the driving dynamics of real vehicles.

The simulation software that runs on the vehicle-in-the-loop is called virtual test drive (VTD), Figure 9.3. VTD consists of different types of components: key components control data flow within the simulation; expanded components are used for simulating vehicles and the environment; while a target component is supplied by the developers of an assistance system.

One of the most important software components for simulating assistance and safety systems is the visualization that depicts the scene graphically in a realistic manner. Visualization output is linked to a head-mounted display that is worn by the driver of the VIL vehicle, Figure 9.4. The display is semi-transparent, so that virtual reality can be projected onto the actual road. For nighttime simulations, it is possible to make only the virtual reality visible to the driver. This has advantages for applying a light function so that it is not necessary to drive exclusively at night. It is a disadvantage that large vacant areas have to be used for tests in this operating

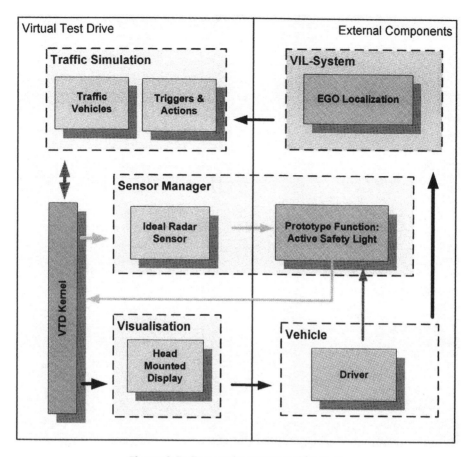

Figure 9.3: Test environment architecture.

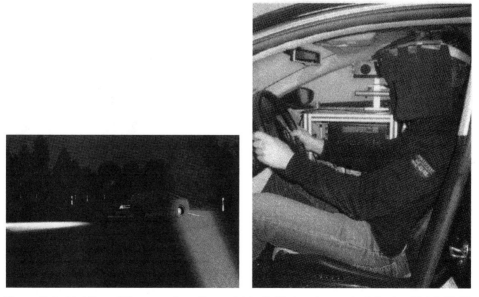

Figure 9.4: Lighting of the rear of another vehicle (left); head-mounted display (right) [9-5].

mode, because driver orientation is exclusively oriented to the virtual reality and cannot establish any reference to the actual environment [9-5].

Testing with the VIL means that driving dynamics do not have to be simulated. Actual driving dynamics are measured in six degrees of freedom using internal sensors including accelerometers, gyroscopes, and differential GPS (DGPS). The data from these and other devices are then incorporated into the VTD simulation in real time. This allows values such as the vehicle position, speeds, and yaw rate to be used to replicate the real world for the driver as realistically as possible. For this reason, the driver's head position is also measured so that the driver can observe the environment as he would in an actual vehicle.

The VIL is compact and mobile. It can be installed in any car of any brand and type, following an uncomplicated and straightforward procedure. It needs only to be connected to the vehicle's power supply and communication system, whether it be CAN, Flexray, or IP-based (Figure 9.5).

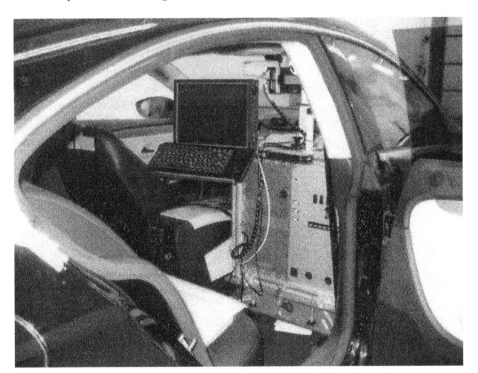

Figure 9.5: VIL simulator.

9.4 Crash simulation

9.4.1 Introduction

Since the beginning of the 1970s computer support for crash simulation has steadily increased. Whereas in the 1980s, supercomputers like the Cray [9-6] were used, powerful single computers are now available in a networked environment. The reason for this evolutionary development is the ever-increasing complexity of the development process and the fact that simulation accuracy must be as high as possible while meeting the demands for speed and low cost. Simulations are used for component development, for a broad range of crash tests, and for accident simulation and reconstruction. During a collision, a shock wave propagates through the vehicle structure with seismic velocity. This phenomenon generates both elastic and plastic deformations. Material properties and characteristics as well as occupant kinematics and dummy behavior must be known and form important parameters for accurate, reliable, and thus, usable simulations.

Most crash configurations are the product of legal requirements, while others are based on consumer tests and tests that have been specifically developed by vehicle manufacturers.

These crash configurations include frontal, lateral, and rear-end collisions; rollovers; car to car collisions; car to pedestrian collisions; repair costs tests; head impact to the interior ejection tests; quasi-static rollover; and head restraint performance tests.

Rigid body simulation (MKS or Multi Body Simulation) was specifically developed to reduce computer time. Using this method, a mechanical system can easily be converted into almost any imaginable vehicle system and vehicle interior and occupants can be simulated. Increased computer power not only makes simulation of vehicle crash performance possible but also the dummy simulation utilizing FEM (Finite Element Methodology).

The following are a few examples of computer simulation used today during the development process.

9.4.2 Frontal crash

The frontal crash includes the following:

- 90 degree test against a fixed barrier
- 30 degree test against a fixed barrier
- offset crash against a deformable barrier
- vehicle to vehicle collisions full frontal or offset
- vehicle to pole collision

Figure 9.6 shows a simulation of an offset crash against a deformable barrier in which not only vehicle crush and performance must be simulated but also the behavior of the deformable barrier and the various sized dummies.

9.4.3 Lateral impacts

For the simulation of lateral impacts, at a minimum the following crashes must be simulated:

- vehicle to vehicle
- deformable barrier to vehicle
- vehicle to pole

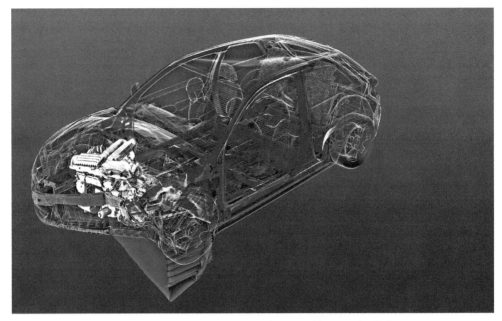

Figure 9.6: FEM Computer simulation, offset crash, [9-7].

The number of parameters required for these simulations increases significantly compared to those required for frontal crash simulations. Figure 9.7 shows a simulation of a side impact using a sub-structure with the SIPsim/PAMCrash simulation program [9-8].

Not only the kinematics but also vehicle performance becomes increasingly realistic as a result of the power and sophistication of hardware and software programs, as evident in Figure 9.8. Here, a computer simulation of a side impact is shown on the left side and is compared to the real side impact on the right.

9.4.4 Rear-end collision

Rear-end collision simulations include vehicle to vehicle and movable barrier to vehicle tests.

The purpose is the simulation is to analyze occupant loading and kinematics, for example, whiplash, as well as fuel-system, fuel tank performance, and high-voltage battery integrity.

9.4.5 Rollover

Quasi-static roof performance and resistance to deformation as well as the kinematics of restrained and unrestrained vehicle occupants are important issues during the vehicle development process.

The containment test requires that a defined head form remain within the confines of the vehicle compartment during the test. The simulation of belted and especially unbelted occupants requires the careful definition of vehicle structure and

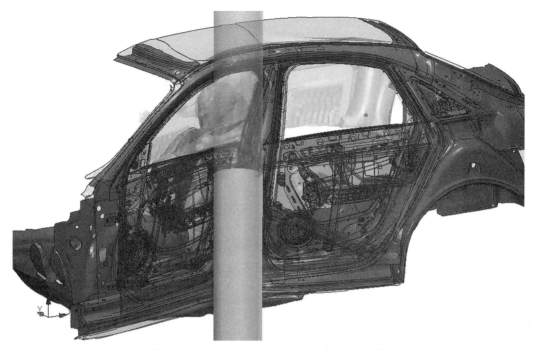

Figure 9.7: Pole side impact simulation [9-8].

the vehicle interior. Interaction between interior components such as the steering wheels and occupants is the focus of many of these simulations. Figure 9.9 demonstrates occupant movement relative to the vehicle as well as possible interactions of occupants with the steering wheel [9-9].

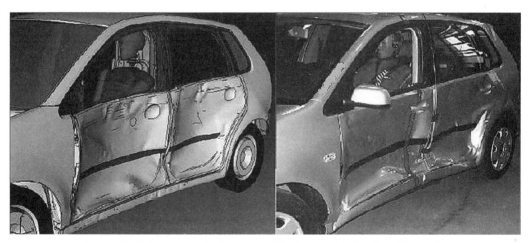

Figure 9.8: Side impact with a movable deformable barrier at 30 mph, left computer simulation, right actual crash test.

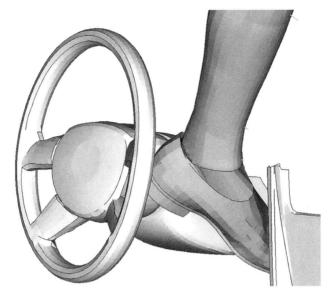

Figure 9.9: Interaction of an unbelted occupant with the steering wheel during a rollover simulation [9-9].

9.4.6 Components

Nearly all safety related components can be simulated and tested including: sensors, bumpers, human kinematics, safety belts, seats, back rests, head restraints, vehicle interiors like the instrument panel and vehicle compartment, deformation zones, air bags, and steering wheels. Figure 9.10 shows a simulation of the head form impacting the vehicle interior [9-10].

Simulation methods such as MADYMO, PAM-Crash, LS-DYNA, ABAQUS, RASIOS and other special programs are used as standard development tools.

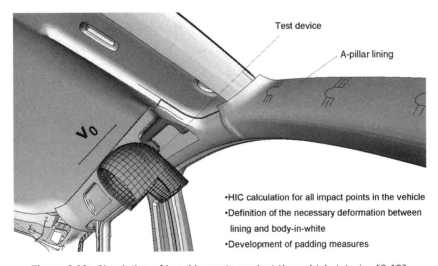

Test device

A-pillar lining

•HIC calculation for all impact points in the vehicle
•Definition of the necessary deformation between lining and body-in-white
•Development of padding measures

Figure 9.10: Simulation of head impacts against the vehicle interior [9-10].

9.5 References

9-1 Lemmer, K. Mensch-Maschine Interaktion. Handbuch Kraftfahrzeugtechnik. Vieweg-Teubner. 2011. ISBN 978-3-8348-1011-3.

9-2 Vollrath, M. Lemmer, K. Wahrnehmung von Assistenzsystemen. Symposium Automatisierungs- und Assistenzsysteme für Transportmittel. AAET 2003, Braunschweig.

9-3 Köster, F. et. al. Entwicklung haptischer Interaktionsmuster zur Führung hoch automatisierter Fahrzeuge über ein Steer-by-Wire-System. AAET 2012. Braunschweig. ISBN 978-3-937655-27-7.

9-4 Bock, Th. Vehicle-in-the-Loop-A new simulator set-up for testing Advanced Driving Assistance Systems. Driving Simulation Conference 2007. Iowa City, IA.

9-5 Gonter, M. et. al. Active safety light-using virtual test drive within vehicle-in-the-loop. IEEE-ICIT 2010. Chile.

9-6 Seiffert, U., Scharnhorst, Th. Die Bedeutung von Berechnungen und Simulationen für denAutomobilbau, VDI Berichte Nr. 699. Würzburg. 1988.

9-7 AUDI AG 2012.

9-8 Elsäßer, K. ADVANCES. TRW Automotive, Issue 33, 2011.

9-9 Borgmann, P. Ph.D. Thesis 02-2012. Technical University Braunschweig.

9-10 Oehlschläger, H. et. al. FEM-Crash simulation: Ein modernes Werkzeug in der Nutzfahrzeugentwicklung. Proceedings of the Verband der Automobileindustrie. (Association of the German Automobile Industry VDA) Technical Congress IAA Germany 2000.

Chapter 10
Looking into the Future

10.1 General trends

We are living in a fast-changing world, where forecasts, especially in areas like safety, are hard to make. In this chapter we try to define some thoughts and observations, which need to be realized over the coming years and some apparent changes that are influencing road traffic safety.

- The population of the world continues to grow.

- The number of megacities is increasing, although the percentage of people who will not be living in megacities will still be relatively high. Whether special cars for the transportation of people we will have in megacities will be needed, how fast they will be needed, and how many will be needed is still unclear.

- There is definitely a market for car sharing. How big this market will be is an interesting question. Aside from consumer acceptance, we must make sure that these cars are properly maintained and kept safe and clean.

- Connected driving, modern information, communication, and driver assistance systems inside and outside the vehicle will connect the driver, the vehicle, and the environment whenever the vehicle is being used [10-1].

- More and more assistance systems will lead to a higher degree of safety. It goes without saying that such systems must be reliable and must not distract the driver from the driving task.

- Gasoline, diesel, and alternative fuels and energies like CNG, LPG, hydrogen, and electricity will all be in the market and will require new powertrain systems, including hybrids, plug-in hybrids, and electric drive with batteries and fuel cells in addition to improved and optimized combustion engines.

- Not only the safety of passenger cars but also the safety of trucks, two-wheelers, pedestrians, and public transportation has to be improved.

- All changes mentioned above will have an effect on traffic safety. But the real success in reducing injuries and fatalities will achieved only if all governmental authorities work together with industry, science, and technology. Education of traffic participants and improvements in the infrastructure and traffic management systems are critical areas for consideration. Traffic safety is a function not just of the vehicle, but also of the driver and the environment including the infrastructure.

10.2 Future of vehicle safety

To further reduce road fatalities and injuries more improvements are necessary in mitigation of injuries or passive safety, accident avoidance or active safety, and integrated safety.

The levels of safety achieved to date in the field of injury mitigation during severe accidents cannot be reduced because of improvements in the field of accident avoidance. The opposite must be the case.

As more detailed information is derived from accident analysis, further optimization of the current safety levels will follow. Two examples for this thesis are the increased roll-over protection, especially in the U.S. and advances in adaptive restraint systems.

Future vehicles will be able to communicate with the environment and provide the driver with the relevant information required to avoid hazardous situations and provide timely information and warnings to all of those involved in the traffic mix.

The number of driver assistance systems will increase, but 100% autonomous driving installed as standard equipment in all cars will not occur in the predictable future.

Presafe systems will be rolled out in a larger number of vehicles and new systems that help to achieve shorter braking distances, as shown in Figure 10.1 may well be feasible.

Special attention will be required because of ongoing changes in powertrain systems and technology, including various alternatives currently under development, for example, the safety of high-voltage systems, fire safety (short cuts, alternative fuels, H_2), and fuel system safety.

A typical example in this area is reversible and irreversible high-voltage shut down and battery discharge response depending on accident severity as shown in Figure 10.2.

The vehicle-to-vehicle, vehicle-to-infrastructure communication, and driver assistance systems will also be able to improve the safety of unprotected traffic participants including pedestrians.

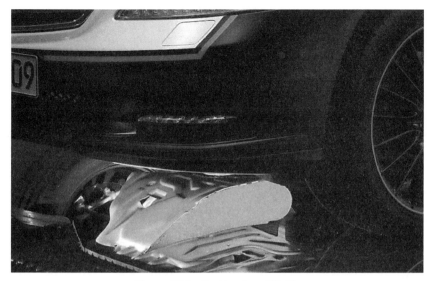

Figure 10.1: Braking bag [10-2].

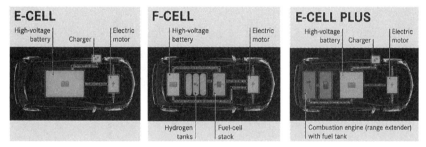

Figure 10.2: Safety features for alternative power train systems [10-2].

10.3 Responsibility of the government authorities

As mentioned above, the goal of reducing the number of worldwide accidents is, to a very great extent, only possible if governments take their responsibilities as seriously as does science and industry. This means that they:

- Support and encourage new technology with appropriate legislation
- Support and encourage research activities
- Develop and maintain a modern and up-to-date infrastructure
- Develop and implement, on an international level, uniform standards for infrastructure that can be used by driver assistance systems to increase operational and traffic safety
- Initiate and promote the worldwide harmonization legislation relating to vehicle and traffic safety including accident investigation and standardization of reporting parameters

- Support and encourage education programs for all type of traffic participants
- Ensure that the security of vehicles and road systems (no external influence on soft and hardware) is guaranteed.

10.4 A final remark

Both of us teach young university students in fields related to vehicle safety. During the lecture we have often observed that automotive safety generates high enthusiasm amoung young students. The motivation of engineers to create engineering solutions to reduce accident risk, to save lives and to promote safety has, without doubt, significantly contributed to the successes achieved to date. Much remains to be done to increase the existing levels of traffic safety in the years to come. This is a responsibility that all of us share.

10.5 References

10-1 Grote, Ch. BMW Connected Drive-Effizienz, Komfort und Sicherheit durch Vernetzung. AAET 2012 ITS –Niedersachsen e.V., Braunschweig. ISBN 978-3-937655-27-7.

10-2 Schöneburg, R. Neue Sicherheitstechnik und Fahrerassistenzsysteme bei Mercedes. HTL-Steyr, Austria, December 1, 2011.

Index

About the Authors

Professor Dr.-Ing. Ulrich Seiffert currently is acting chairman of WiTech Engineering GmbH, member of the board of ITS Niedersachsen, member of the board of VDI Vehicle Technology, member of the Helmholtz Senat, member of the Royal Swedish Academy of Engineering Sciences (IVA) and acatech.

Until the end of 1995, Dr. Seiffert was a member of the board for research and development of Volkswagen AG in Germany. He is a board member of several industry, research and social organizations. He also holds a number of patents in the safety area and for vehicle and traffic technology and has received awards for his work in the field of safety and environmental engineering. He is the author of a large number of books and has published more than 350 technical papers.

Dr. -Ing. Mark Gonter, head of integrated safety and light in the group research at Volkswagen, has been professor of integrated vehicle safety at the University of Braunschweig, Germany, since 2008. Furthermore, he is a lecturer at the Autouni of Volkswagen. Dr. Gonter has worked for the Volkswagen Group since 1999. His previous roles include head of integrated safety, project manager of adaptive occupant safety, and accident researcher. In 2007 he received a doctorate of engineering in intelligent restraint systems from Dresden University of Technology in Germany. He has published a large number of technical papers in the field of active and passive safety and holds a number of patents that pertain to automotive vehicle safety.